Targeted Regulatory Writing Techniques: Clinical Documents for Drugs and Biologics

Edited by Linda Fossati Wood and MaryAnn Foote

Birkhäuser
Basel · Boston · Berlin

Linda Fossati Wood
Medwrite Inc.
Westford, MA 01886
USA

MaryAnn Foote
MA Foote Associates
Westlake Village, CA 91362
USA

Library of Congress Control Number: 2008929030

Bibliographic information published by Die Deutsche Bibliothek
Die Deutsche Bibliothek lists this publication in the Deutsche Nationalbibliografie;
detailed bibliographic data are available in the Internet at <http://dnb.ddb.de>.

ISBN 978-3-7643-8361-9 Birkhäuser Verlag, Basel – Boston – Berlin

© 2009 Birkhäuser Verlag, P.O. Box 133, CH-4010 Basel, Switzerland
Part of Springer Science+Business Media
Printed on acid-free paper produced from chlorine-free pulp. TCF ∞
Printed in Germany
Cover illustration: Bradley Poff, Medford MA, USA
Cover design: Alexander Faust, Basel, Switzerland
ISBN 978-3-7643-8361-9   e-ISBN 978-3-7643-8362-2

9 8 7 6 5 4 3 2 1   www.birkhauser.ch

Contents

List of contributors

Peggy Boe, Image Solutions, Inc., Wilmington, North Carolina, USA;
e-mail: peggy.boe@imagesolutions.com

Jennifer A Fissekis, Rye Brook, New York, USA; e-mail: jfisseki@optonline.net

MaryAnn Foote, MA Foote Associates, Westlake Village, California, USA;
e-mail: fmawriter@aol.com

Takumi Ishida, Japan Medical Linguistics Institute, Kobe, Japan;
e-mail: uih37469@nifty.com

Katsunori Kurusu, Marketed Products Regulatory Affairs, Sanofi-Aventis KK,
Tokyo, Japan; e-mail: katsunori.kurusu@sanofi-aventis.com

Debra L Rood, Liquent Certified Advanced Publisher, Simi Valley, California,
USA; e-mail: debrarood@yahoo.com

Linda Fossati Wood, MedWrite, Inc., Westford, Massachusetts, USA;
e-mail: lfwood@comcast.net

James Yuen, Editorial Consultant, Moorpark, California, USA;
e-mail: james.yuen@alum.calberkeley.org

List of reviewers

Justina A Molzon, Associate Director for International Programs, US Food and Drug Administration, Center for Drug Evaluation and Research, Silver Spring, Maryland, USA

Lisa Quaglia, Director Regulatory Affairs & Clinical Research, Straumann USA, LLC; Andover, Massachusetts, USA

Donald Smith, Director Global Regulatory Writing, Amgen Inc; Thousand Oaks, California, USA

James Yuen, Editorial Consultant, Moorpark, California, USA

Foreword

This book describes the authors' standard or 'best' practices used in writing regulated clinical documents for the drug and biologics industry. The fundamental premise of this book is that the end (documents submitted to a health authority) is dependent on the beginning (the planning and strategy that go into organizing written documentation). Each regulatory document inherently exists within a constellation of related documents. This book attempts to show the relationships between and among these documents and suggests strategies for organizing and writing these documents to maximize efficiency while developing clear and concise text. At all times, and irrespective of applicable laws and guidelines, good communication skills and a sense of balance are essential to adequately, accurately, and clearly describe a product's characteristics. At no time should the reader perceive these suggestions to be the only viable solution to writing regulatory documents nor should the reader expect that these suggestions guarantee product success.

The audience for this book is the novice medical writer, or those who would like to explore or enhance regulatory-writing skills. We assume the reader will have a basic understanding of written communication, but little experience in applying this skill to the task of regulatory writing. Extensive knowledge of science, clinical medicine, mathematics, or regulatory affairs law is not required to use the best practices described in this book.

The scope of this book is regulatory writing of clinical documents and clinical sections of regulatory submissions for drugs and biologics during premarketing stages of product development. This type of writing is described within the context of a regulated environment for Europe, Japan, and the United States. Because the editors and chapter authors are most experienced with writing documents for the United States regulatory authorities, these documents are the primary focus of this book. The exception is Chapter 12 (Clinical trial procedures and approval processes in Japan), with a focus on the regulatory requirements in Japan. Many other regions of the world also require regulated clinical documents but discussion is not within the scope of this book.

Regulatory writing techniques also are used for medical devices, for nonclinical and manufacturing writing, and during the postmarketing phase of development, but these documents are outside the scope of this book. The list of documents included here is meant to represent those documents that are most frequently written by a regulatory writer. The list is by no means exhaustive, as many additional documents may be required based on product-specific characteristics or global region.

It should be noted that the opinions expressed by chapter authors may not necessarily reflect the opinions of the editors. We have taken due diligence to ensure that all information is current and correct, but we are not responsible for errors, omissions, or commissions. Discussion of a product is not endorsement for its use.

We hope that you enjoy the book and that it helps you in clarifying your thinking as you prepare your regulatory submissions.

April 2008

Linda Fossati Wood, RN, MPH
Westford, Massachusetts

MaryAnn Foote, PhD
Westlake Village, California

Regulatory writing fundamentals

Targeted Regulatory Writing Techniques. Clinical Documents for Drugs and Biologics,
edited by Linda Fossati Wood and MaryAnn Foote
© 2009 Birkhäuser Verlag Basel/Switzerland

Chapter 1

Developing a target

Linda Fossati Wood

MedWrite, Inc., Westford, Massachusetts, USA

Introduction

> *Finis origine pendet* (The end depends on the beginning)
> Attributed to Roman poet Manlius

Regulatory writing is an integral part of the health-product development process. Most nations have a governmental authority (also called a regulatory agency) responsible for determining whether a drug or biologic is sufficiently safe to allow commercial distribution. The product's manufacturer must provide written documentation to this regulatory agency (called a submission) making an argument for safety and efficacy of the product. The regulatory agency, if it approves of the data and the claims, will file the submission and grant marketing approval. Regulatory writing is the discipline responsible for development of these regulatory documents.

Regulatory writing is important to companies that wish to market and sell their healthcare products and also is important to the general public that uses these products. Clear, concise text that communicates corporate goals and satisfies local and international regulatory requirements is critical to successful and rapid product approval for commercial distribution. Most importantly, an accurate and clear characterization of a product's safety and efficacy is an essential part of medical care.

Standard methods, also called 'best practices,' have been used by the authors of this book to write regulated clinical documents for the drug and biologics industry. The point of these best practices is to plan for the end (documents submitted to a health authority), by developing a document strategy at the beginning. The authors attempt to show the relationships between and among these documents, and they suggest strategies for organizing and writing these documents to maximize efficiency while developing clear and concise text.

Best practices in regulatory writing are described in terms of five tasks:

- Developing a target: Determining which document(s) is needed based on five steps: classification of the product, the geographic region in which the product will be marketed, the stage of development, the intended content, and bringing these 4 steps together to determine the document(s) to be written (Chapter 1, Developing a target).
- Using a writing toolkit: Selecting and using general principles of regulatory writing (Chapter 2, Regulatory writing tips); templates and styles (Chapter 3, Templates and style guides); and developing procedures for document review (Chapter 4, Document review).
- Writing source documents: Writing the documents that form the basis for all integrated documents and submissions (Chapter 5, Protocols; Chapter 6, Clinical study reports).
- Writing integrated documents: Writing documents that integrate and summarize information from source documents (Chapter 7, Investigator's brochures; Chapter 8, Investigational medicinal products dossier; Chapter 9, Integrated summaries of safety and efficacy; Chapter 10, Informed consent forms).
- Writing submissions: Putting the source and integrated documents together (Chapter 11, Global submissions: The common technical document; Chapter 12, Clinical trial procedures and approval processes in Japan; Chapter 13, Region-specific submissions: United States of America).

Unlike many types of writing, regulatory writing is not a solitary task. All regulated documents described in this book are the result of collaboration with a team and as such reflect the cross-disciplinary efforts and expertise of the team members. The specific functional areas included on each development team vary by company and document, and occasionally by product. We suggest that team members should be included during development, with the caveat that not all are always required for each area and the best teams may be flexible, comprising members from additional functional areas .

The first step in regulatory writing is to ascertain which document needs to be written and should be determined in collaboration with clinical and regulatory staff. The writer should have sufficient knowledge to understand the context within which the document will be written. Determining the document to be written requires categorization of products using the following steps:

- Step 1: Product classification: Is it a drug, biologic, medical device, or combination product?
- Step 2: Geographic region: Will the application be submitted in Europe, Japan, or the United States, the three major regions that drive regulatory documentation? Or will it be submitted to another region of the world?
- Step 3: Stage of product development: Is the product currently being sold (also called marketed) or is it in premarketing development?
- Step 4: Source or integrated document: How many studies are being described? A source document describes one study, an integrated document describes more than one study (often with an integrated analysis of data across two or more studies) or may cross company departments.
- Step 5: Developing a target: using information from the first four steps, the document(s) required is evident.

Side bar: Lessons learned

It is impossible to overstate the importance of this type of rudimentary planning, which intuitively would be the logical first step when embarking on a project with such scope and impact. The editors sadly can attest to problems encountered when upfront planning for a regulatory submission was inadequate. While many submission team members may balk at the time spent in planning what documents are needed, who will write each document, how documents will be reviewed and changes agreed on, and other planning details, experience has shown us that detailed planning saves time. The maxim is every day off market for a good product is a loss of US$1 million; this statistic alone should bolster the writer's (and the team's) efforts for planning.

Step 1: Product classification

Although regulatory writers are not responsible for determining whether an investigational product is a drug, biologic, or medical device, an understanding of the distinction between drugs and biologics and medical devices is important because of the difference in documents.

Drugs
Drugs (also called pharmaceuticals) are chemical entities that affect metabolism. The European Medicines Agency (EMEA) in Europe, the Ministry of Health, Labour and Welfare (MHLW) in Japan, and the United States Food and Drug Administration (FDA) Center for Drug Evaluation and Research (CDER) regulate drug

testing, manufacturing, and sales. The United States Food, Drug & Cosmetic Act (FD&C Act) defines drugs by their intended use:

- Articles intended for use in the diagnosis, cure, mitigation, treatment, or prevention of disease, and
- Articles (other than food) intended to affect the structure or any function of the body of man or other animals [1].

Biologics

Biologics, in contrast to drugs that are chemically synthesized, are derived from living sources (such as humans, animals, and microorganisms) [2]. The EMEA in Europe, the FDA Center for Biologics Evaluation and Research (CBER) in the United States, and the MHLW in Japan regulate the companies that test, manufacture, and sell biologic products.

The United States Code of Federal Regulations (CFR) defines a biologic product as any virus, therapeutic serum, toxin, antitoxin, or analogous product applicable to the prevention, treatment, or cure of diseases or injuries of humans [3]. Regulation of biologics is similar to that of drugs, so documentation of clinical research and development generally follows the drug model. We describe documentation for drugs and biologics together.

Medical devices

Although writing for medical devices is beyond the scope of this book, a few basic principles of medical device development will be explained to differentiate these products from drugs and biologics.

Medical devices range from simple tongue depressors and bedpans to complex programmable pacemakers with microchip technology and laser surgical devices. If the primary intended use of the product is not achieved through chemical action or metabolism by the body, the product is usually considered to be a medical device [4].

The European Commission (EC) in Europe, MHLW in Japan, and the United States FDA's Center for Devices and Radiological Health (CDRH) are responsible for regulating firms that test, manufacture, and sell medical devices. In addition, CDRH regulates radiation emitting electronic products (medical and nonmedical) such as lasers, radiographic (x-ray) systems, ultrasound equipment, and microwave ovens [5].

Under the European Union's (EU) Medical Device Directive, a medical device is defined as any instrument, apparatus, appliance, material, or other article, whether used alone or in combination, including software necessary for its proper application intended by the manufacturer to be used for humans for the purpose of:

- Diagnosis, prevention, monitoring, treatment, or alleviation of disease,
- Diagnosis, monitoring, treatment, alleviation of, or compensation for an injury or handicap,
- Investigation, replacement, or modification of the anatomy or of a physiological process, or
- Control of conception

and which does not achieve its principal intended action in or on the human body by pharmacologic, immunologic, or metabolic means, but which may be assisted in its function by such means [6].

Device regulations differ greatly from those applied to drugs and biologics by virtue of stratifying devices into several classes that determine the degree of rigor required for approval to sell the device. The system of classification established by the EU, Japan, and the United States differ somewhat, but all attempt to quantify the degree of 'risk' posed by the device.

European Union's classification
Device classification is defined in the Medical Device Directive and is based on a complex set of rules that define device risk by duration of use and invasive character-istics [6]. Classifications range from Class I (lowest risk) to Class III (highest risk).

Japan's classification
Japan's system of medical device classification is based on level of risk, which deter-mines whether clinical information is required [7].

- Class I: Clinical data not required.
- Classes II–IV: Ranges from relatively low risk (no clinical data required) to pos-sible fatal risk in case of failure (clinical data required).

United State's classification
The system used in the United States considers three classes [4]:

- *Class I general controls:* Class I devices are the lowest risk devices and generally do not require FDA notification or approval before sales and distribution
- *Class II general controls and special controls:* 510(k) Premarket Notification is re-quired before commercial distribution. The submission makes the argument that the device is "substantially equivalent" to another device legally marketed in the United States before May 28, 1976, or to a device that has been determined by FDA to be substantially equivalent. The 510(k) is notification and does not require approval from FDA before commercial distribution, but it does require FDA con-currence that the device is "substantially equivalent" to a legally marketed predi-cate device before commercialization.

- *Class III general controls and premarket approval:* A Premarket Approval (PMA) Application is required before commercial distribution for most Class III medical devices. In general, products requiring a PMA are high-risk devices (life-saving, life-sustaining, or breakthrough technology) that pose a significant risk of illness or injury. The PMA process is more involved than the 510(k) process and includes the submission of clinical data to support claims made for the device. The PMA is an actual approval of the device by FDA.

Combination products

The term 'combination product' includes a product that comprises [8]:

- Two or more regulated components (ie, drug/device, biologic/device, drug/biologic, or drug/device/biologic) that are physically, chemically, or otherwise combined or mixed and produced as a single entity;
- Two or more separate products packaged together in a single package or as a unit and composed of drug and device products, device and biologic products, or biologic and drug products;
- A drug, device, or biologic product packaged separately that, according to its investigational plan or proposed labeling, is intended for use only with an approved individually specified drug, device, or biologic product where both are required to achieve the intended use, indication, or effect and where upon approval of the proposed product the labeling of the approved product would need to be changed (eg, to reflect a change in intended use, dosage form, strength, route of administration, or significant change in dose); or
- Any investigational drug, device, or biologic product packaged separately that, according to its proposed labeling, is for use only with another individually specified investigational drug, device, or biologic product where both are required to achieve the intended use, indication, or effect.

Regulatory writing for combination products poses its own set of challenges, as the written documents must be modified from those required for each of the component products (drug/device, biologic/device, drug/biologic, or drug/device/biologic). As defined regulations or guidelines for combination products are still in their infancy in development, a best practice for writing clinical documents is to use the product classification with the most rigorous regulatory definition. This best practice generally means that combination products comprising medical devices will be written as for a drug product. The extensive, exhaustive, and at times, excessive, level of detail required for description of a drug product, however, may not be appropriate for a medical device, even a device that is under development as a combination product. Good communication skills and a sense of balance are important to determine the level of detail required.

Step 2: Regions of the world

After ascertaining the product's classification, the second step in developing a target is to identify the region in which the product will be tested and commercially distributed, as this is essential to determining the types of documentation required. The decision to submit in a particular region reflects corporate goals and is not within the regulatory writer's purview; however, the writer needs to be clear on the intended region for submission, as this may influence the documents required.

Three major regions of the world drive the regulatory environment for medical products: the EU, Japan, and the United States. Each of these three regions has a branch of government with authority over regulation of these products and individual regulations for the purpose of controlling the quality of medical products available for commercial use (Table 1). Writing documents for regions other than the major three regions requires close collaboration with staff in Regulatory Affairs. Company experience and negotiations with the health authorities should help guide the writer.

Table 1. Global regulatory authorities and regulatory initiatives by product classification

Geographic region	Drugs/Biologics	Medical devices
European Union (EU)		
Regulatory Authority	European Medicines Agency (EMEA)	Notified Bodies (NB) Competent Authorities
Regulatory Initiative	International Conference on Harmonisation (ICH)	Global Harmonization Task Force (GHTF)
Japan		
Regulatory Authority	Ministry of Health, Labor, and Welfare (MHLW): Pharmaceuticals and Medical Devices Agency (PMDA)	
Regulatory Initiative	International Conference on Harmonisation (ICH)	Global Harmonization Task Force (GHTF)
United States of America		
Regulatory Authority	Food and Drug Administration (FDA)	
	Center for Drug Evaluation and Research (CDER) Center for Biologics Evaluation and Research (CBER)	Center for Devices and Radiological Health (CDRH)
Regulatory Initiative	International Conference on Harmonisation (ICH)	Global Harmonization Task Force (GHTF)

The EMEA, which began its activities in 1995, coordinates the evaluation and supervision of medicinal products throughout the 27 member nations of the EU [9]. Medical devices in the EU are regulated by the EC, which has issued the Medical Device Directives [6].

The Japanese MHLW regulates drugs, biologics, and medical devices under the Pharmaceutical Affairs Law (PAL; Law No. 145 issued in 1960) of the Pharmaceutical and Medicinal Safety Bureau (PMSB) [10]. This legislation describes the requirement for Clinical Trial Notification (CTN) and Marketing Approval Application (MAA). The CTN and MAA are submitted to the MHLW and then reviewed by an Independent Administrative Institution, the Pharmaceuticals and Medical Devices Agency (PMDA). MHLW has the authority to approve drugs for testing in humans, and for marketing and distribution (Chapter 12, Clinical trial procedures and approval processes in Japan).

Regulation of drugs, biologics, medical devices, and combination products is the responsibility of the FDA in the United States. The FDA is an agency within the Department of Health and Human Services, and consists of eight centers [11], three of which are important to understanding regulatory writing of clinical material for healthcare products:

- Center for Drug Evaluation and Research (CDER);
- Center for Biologics Evaluation and Research (CBER); and
- Center for Devices and Radiological Health (CDRH).

The EMEA, MHLW, and FDA define the documentation required for testing and commercialization in their respective regions.

In addition, several regulatory initiatives have been formed that affect written documents for all of these regions (Table 1). These efforts are represented by the International Conference on Harmonisation of the Technical Requirements for Registration of Pharmaceuticals for Human Use (ICH, drugs and biologics) and the Global Harmonization Task Force (GHTF, medical devices). The purpose of initiatives such as the ICH and the GHTF is to bring harmonization, that is, consistency in requirements to product development. Securing the right to sell a product requires that the product's manufacturer sends (or submits) a group of documents to one or more regulatory agencies. The requirements for all regions differ, sometimes substantially, so effectively securing approval for selling a product in different geographic regions of the world has traditionally been a daunting, time-consuming, and expensive task. Hence, efforts at harmonization, or aligning requirements across regions, have been initiated for both drugs and medical devices.

The ICH, created in 1990, is an agreement among the EU, Japan, and the United States to harmonize different regional requirements for registration of pharmaceutical drug products [12]. Such a joint effort by regulators, the biopharmaceutical

---

Side bar: Lessons learned

Many global companies have regulatory writers based in Europe, Japan, and the United States who can answer questions and provide documentation to regulatory agencies during their normal business hours when counterpart offices are closed. If this model of regulatory writing is used, it is useful, particularly in the beginning and if any managers are hired, for writers to spend some time in the other offices to learn processes and procedures and to develop some interpersonal relationships.

Because submissions are generally global, it is often useful to have some process by which regulatory documents can be worked on by writers at different times of the day, almost maintaining 24-hour/day work. The lead writer for the project would have final responsibility for overall style and quality, but experience suggests that in a global submission setting that allows the European office access to the document when the United States staff is not in the office, thus makes it possible to meet very tight timelines. Such a process also allows the regulatory writers to have a strategic global role in the submission process.

Document management processes and templates should be standardized across regions and changes suggested, discussed, and agreed to by all writing groups (and any other functional group charged with input, such as statistics). The concept of 'one document, many uses' can speed writing and reviewing time, and document management systems. Chapter 3 discusses standardized templates and boilerplate language.

---

industry, and trade associations is unique, and the working groups have generated a number of guidelines that drive regulatory writing.

Medical devices are not currently included in the ICH guidelines; however, the GHTF is a similar initiative that may eventually bring the various device regulations together. The GHTF was conceived in 1992 in an effort to respond to the growing need for international harmonization in the regulation of medical devices. It is a voluntary group of representatives from national medical device regulatory authorities and the regulatory industry. The GHTF has representatives from five founding members grouped into three geographical areas: Europe, Asia-Pacific, and North America. The primary function of the GHTF is publication and dissemination of harmonized guidance documents on basic regulatory practices [13].

Regulatory initiatives function to put forth guidances (also called guidelines). In contrast to regulations (which are laws), guidances are nonbinding recommendations. Because these guidances provide expanded and helpful interpretations of the regulations, they are very beneficial to the regulatory writer.

Step 3: Stages of product development

The third step in developing a writing target is to ascertain the stage of development as it relates to the ability to market the product. All new drugs and biologics, irrespective of geographic region, follow the same basic, orderly, and highly regulated process of development. Knowledge of the product development process is essential to determining the regulatory documents required at each stage, and these documents vary by geographic region.

For all three geographic regions, the process used comprises: discovery (also called laboratory or bench testing, and consists of in vitro testing of tissues, plasma, etc); nonclinical testing in live animals (in vivo testing); request for permission to test in humans; and testing in humans (Figure 1). These steps are followed by a request for approval to market the product. Each of these stages is associated with specific regulatory documentation.

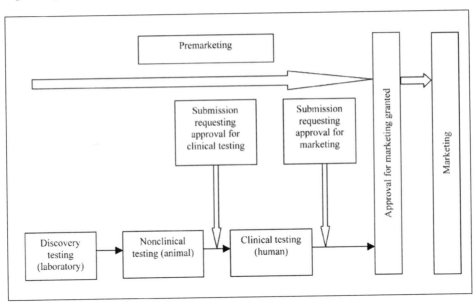

Figure 1. Approval process for drugs and biologics

Before use in humans
During the discovery (or bench) stage, before testing in live animals, a minimum of regulatory writing occurs. The protocols used are brief and reports generally consist of a few pages of text with data sheets appended and an occasional publication. Moving from this stage to nonclinical testing in animals is simple in regulatory terms, as notification of health authorities is not generally required.

Regulatory writing as a function generally starts to become an essential part of product development when animal testing begins. Nonclinical documents are similar to those written for human testing, in that study conduct is planned by the protocol, and results of testing are described in a study report. Documentation of nonclinical studies is beyond the scope of this book and readers are advised to consult other sources for further information.

Request for permission to use in humans

After testing in animals is considered adequate to ensure safe testing in humans, and before initiating human trials, the sponsor must send an assembly of documents called a submission to the health authority in the region of interest. This submission differs based on region (Table 2). After submitting these documents and waiting the region-specific time period, and in the absence of an objection by the regulatory authority, the company may begin clinical trials.

Table 2. Submissions required for use in humans by geographic region

Region	Submission
Europe	Clinical Trial Authorisation (CTA)
Japan	Clinical Trial Notification (CTN)
United States of America	Investigational New Drug Application (IND)

Clinical testing

An understanding of the phases of clinical development is important as it determines the documents required. Phase 1 clinical trials establish the preliminary safety risks for the drug, and often explore pharmacokinetics and pharmacodynamic markers. Because no drug or biologic is without toxicity, a risk:benefit profile must be established so that healthcare professionals and subjects can determine if the drug is suitable for them. Phase 1 trials also establish dose, frequency of administration, route of administration, and use with concomitant drugs and food. Phase 1 trials for drugs are usually conducted in a small number (10–30) of healthy volunteers (ie, people who are free from conditions that could complicate interpretation of data). These subjects are monitored closely at frequent time points using a large number of assessments.

Drugs that are known to have potential serious effects, drugs intended for an indication that would not benefit from testing in healthy volunteers, and biologics are generally tested in subjects with the disease. In biologics, healthy volunteers generally are not used for testing because biologics are proteins that could induce antibody production with potential adverse effects. Sometimes the very first trial, often called 'first in man' or, more properly, 'first in human' is called a phase 1a trial.

The quantity of the first dose of a particular drug administered to humans is based on observations from nonclinical toxicology studies [14]. The no-observed-adverse-effect level (NOAEL, the highest dose of the drug that does not produce a significant increase in adverse effects compared with the control group) of the drug is determined based on three criteria: overt toxicity such as clinical signs, surrogate markers of toxicity such as abnormalities in blood values, and exaggerated pharmacodynamic effects.

The NOAEL is used to calculate the human equivalent dose (HED), using mathematical methods to extrapolate the dose from animals to humans, generally based on body surface area. The selected first dose is administered to a small group of subjects (and can be as few as three subjects), and these subjects are observed for signs of toxicity for a specified period of time. Subsequent increases in the dose (called dose escalation) occur until the maximum tolerated dose (MTD) is reached. Often in phase 1 trials, serum drug monitoring is done to obtain important pharmacokinetic data, including maximum concentration in the serum and the time to maximum concentration.

Phase 2 clinical trials are designed to further explore safety of the investigational drug, to provide early data about efficacy, and provide enough data to design phase 3 trials to confirm the product's safety and efficacy. These trials have a larger sample size than phase 1 trials (generally 30–100), and the frequency and types of assessments are fewer than in phase 1. The larger sample size is intended to improve the probability that statistical analyses will be able to determine a difference between test and control groups, and therefore support the study hypothesis. Although a placebo-controlled trial would yield the best definitive answer, some investigators and regulatory authorities believe that it is unethical to withhold active treatment for some diseases. In such situations, an active control (ie, current therapies considered to be standard of care) might be used instead of a placebo.

The function of phase 2 trials is to help design successful phase 3 trials, but many drugs fail at the phase 2 stage and clinical development is terminated. Failure may have been due to a poor risk:benefit ratio (the risk of using the drug outweighs the possible benefits), poor study design, the wrong endpoint, or a lack of statistical power sufficient to show the difference between the drug and placebo or active control.

Data from phase 3 trials confirm the efficacy of a drug and further characterize the safety of the drug. Phase 3 trials have a large sample size (sometimes in the thousands), and the study designs have inclusion and exclusion criteria, time points, and assessments that tend to mimic standard medical care. The design of a phase 3 trial is crucial because the label for the drug and the marketing claims will be developed on the basis of the results of the assessments.

Request permission to market
After completion of the clinical trials, each of the geographic regions requires a submission, which requests marketing approval. Table 3 presents a list of these submissions by geographic region.

Table 3. Submissions required for marketing by geographic region

Region	Submission
Europe	Marketing Authorisation Application (MAA) Common Technical Document (CTD)
Japan	Marketing Approval Application (MAA) Common Technical Document (CTD)
United States of America	New Drug Application (NDA) Common Technical Document (CTD)

Postmarketing approval

Postmarketing clinical trials are often called phase 4 trials (or even phase 3b trials). Phase 4 trials are designed to add more data to the drug's profile: risks, benefits, and potential use in other disease settings. Phase 4 trials are important to supplement additional requirements from regulatory agencies. Sometimes marketing approval will be granted for a product with the stipulation that phase 4 work will be done within a given time frame. Although hundreds to thousands of people can be studied in phase 3 trials, it is often not possible to predict potential side effects in a large, heterogeneous population.

Postmarketing commitments made between drug sponsors and regulatory agencies often include studies in special populations, such as infants, young children, adolescents, the elderly, or subjects with liver or kidney impairment. Other phase 4 commitments may include studies to provide further information about drug-drug interactions, particularly if the drug will be used by a population with co-morbidities that also require drug therapy. Phase 1, 2, 3, and 4 studies are summarized in Table 4.

Table 4. Summary of phase 1, 2, 3, and 4 clinical trials

	Phase 1	Phase 2	Phase 3	Phase 4
Outcome	Safety Dose finding Pharmacokinetic profile Pharmacodynamic markers	Safety Preliminary efficacy Response rate	Safety Efficacy Survival	Safety Efficacy Survival
Participants	Healthy volunteers Subjects with no other treatment options Usually <30	Subjects with the target disease Usually 30 to 100	Subjects with the target disease Usually >100	Subjects with the target disease Often >1000
Drug dose and schedule	Often escalating dose on a fixed schedule	Usually a fixed dose on a fixed schedule	Fixed dose on fixed schedule	Marketed dose and schedule

Step 4: Source versus integrated documents

Writing regulatory documents requires an understanding of the concept of source versus integrated documents, as information is methodically 'built,' starting with source documents, then using these to build integrated documents and submissions.

Source documents

Source documents describe a single clinical study. The description may be either the plan for conduct of the study (a clinical protocol), or results of testing after the study is completed and all data have been analyzed (a clinical study report). A 'study' is defined by the protocol under which the study is conducted (Chapter 5, Protocols) and is limited to the results of testing in the people who were enrolled in that study. These results are described in a clinical study report (Chapter 6, Clinical study reports). Table 5 presents a list of source documents described in this book.

Table 5. Source documents

Source documents	Purpose
Protocol	Describes the objective(s), design, methodology, statistical considerations, and organization of a clinical trial.
Clinical study report	Full report of an individual clinical study, in which clinical and statistical descriptions and presentations are integrated into a single report.

The clinical study report was formerly referred to as an Integrated Clinical Statistical Report in US regulations, in reference to the combination of clinical interpretations and statistical analyses residing in one document.

The information in source documents is used to 'build' integrated documents. Therefore, source documents should be written first, as the accumulated information from more than one source document is used to write integrated documents. As described in Chapters 5 and 6, source documents are specifically organized so that sections of the document may be easily used in other documents. Both of the source documents described in this book (clinical protocols and study reports) have a synopsis and a body of text. The synopsis has the same information as the body of text, but in a summarized form.

The synopsis of a clinical study report is an important piece of information to a regulatory writer. The value of a well-written clinical study report synopsis cannot be overstated, as these pieces of text are used repeatedly in integrated documents and in all premarketing submissions.

Integrated documents

Integrated documents describe more than one clinical study. Integrated documents are intended to provide a summary of product characteristics across studies and may be 'stand-alone' documents such as an investigator's brochure or package insert, or part of submission text such as the clinical overview and clinical summary of the CTD. Table 6 presents a list of integrated documents described in this book.

Table 6. Integrated documents

Integrated document	Purpose
Investigator's Brochure	Compilation of clinical and nonclinical data, and a product description that is relevant to the study of an investigational product [15]
Investigational Medicinal Product Dossier (IMPD)	Compilation of clinical and nonclinical data, summarization of all risks and potential benefits [16]
Informed Consent Form	Documented process by which a subject voluntarily confirms his or her willingness to participate in a particular trial [17]
Integrated Summary of Safety (ISS)	Summary of all safety outcomes for all clinical studies conducted on the product [18]
Integrated Summary of Efficacy (ISE)	Summary of all efficacy outcomes for all clinical studies conducted on the product [18]
Clinical Overview of the Common Technical Document (CTD)	Critical analysis of clinical data, including efficacy and safety outcomes [19]
Clinical Summary of the CTD	Brief, factual summarization of all clinical information [19]

The product discussions in an integrated document require some sort of 'synthesis' of information from more than one study. Descriptions of this synthesis are relatively easy in a small clinical program (two to three studies), but the difficulty escalates quickly with larger programs.

Description of this synthesis is most easily accomplished by the writer if data from all studies have been combined into one database and statistical tables are available. These statistical tables would represent all outcomes studied (age, sex, race, blood pressure, laboratory parameters, etc) presented for each individual study, and for all studies combined.

On occasion (generally because of a small clinical program or a very tight budget) the writer may not be provided with statistical tables representing this combined database. In this case, manual development of tables across studies may be required.

Irrespective of the manner in which the writer receives data from which to write, integrated documents tend to be inherently difficult to write and are not generally under the purview of a novice writer. Effectively describing the safety and efficacy of a product, with an eye to corporate goals and the potential for commercial distribution, tends to require a fair amount of experience.

Relationship of source to integrated documents

Source documents form the basis for integrated documents and should be written before integrated documents. Figure 2 presents the relationship of three source documents (clinical study reports) to a hypothetical integrated document. The source documents contain detailed information relevant to the respective studies (data listings, statistical tables, and the associated text and synopsis). The integrated document contains information from all three clinical study reports.

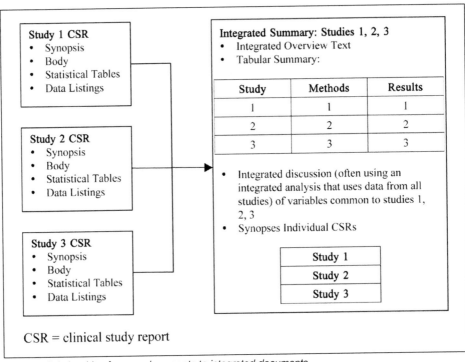

Figure 2. Relationship of source documents to integrated documents

The purpose of writing source documents and integrated documents is to sequentially and methodically distill key messages in support of a product, to reach the point of product labeling. Figure 3 presents the hierarchy of information as it flows from source documents to integrated documents.

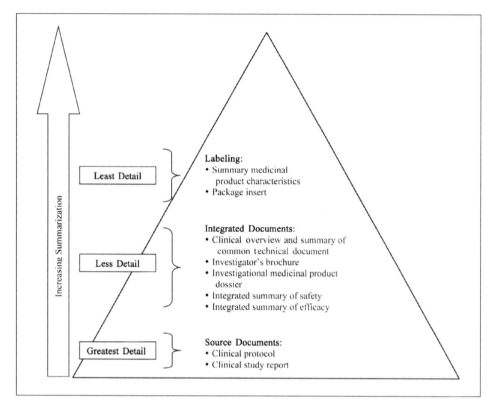

Figure 3. Hierarchy of summarization

Step 5: Determining the target document

As discussed, determining the document(s) to be written (the target) requires knowledge of product classification, region in which the submission is planned, stage of development, and whether the document is a source or an integrated document.

By eliciting this information, the clinical documents required for a specific product (ie, the target of the writing task) will be evident. Figures 4–6 present the three major regions of the world and show the associated stages of development and the clinical documents and submissions required for drugs and biologics.

The reader should note that the list of documents in these figures is by no means exhaustive. A full submission would also contain nonclinical (in vitro and in vivo animal testing) and quality (chemistry, manufacturing, and controls) documentation, and may also contain additional clinical information. The intent is to show those documents most commonly included, and generally (although not always) written by regulatory writers.

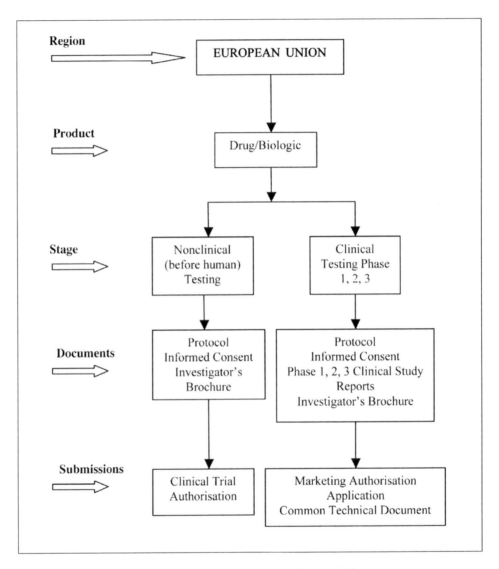

Region → EUROPEAN UNION

Product → Drug/Biologic

Stage →
Nonclinical (before human) Testing
Clinical Testing Phase 1, 2, 3

Documents →
Protocol
Informed Consent
Investigator's Brochure

Protocol
Informed Consent
Phase 1, 2, 3 Clinical Study Reports
Investigator's Brochure

Submissions →
Clinical Trial Authorisation

Marketing Authorisation Application
Common Technical Document

Figure 4. Flow chart for determination of required documents: European Union

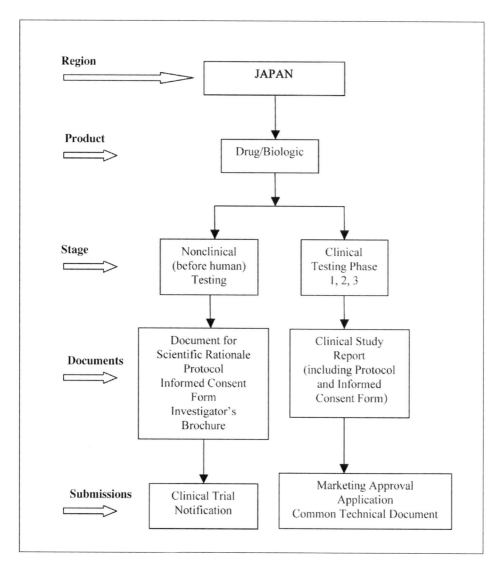

Figure 5. Flow chart for determination of required documents: Japan

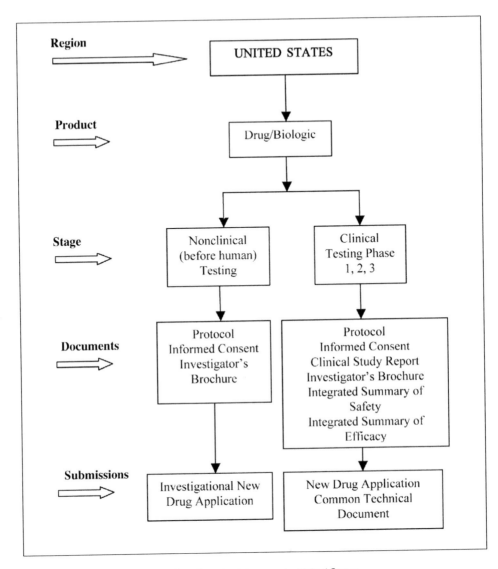

Figure 6. Flow chart for determination of required documents: United States

References

1　Federal Food, Drug, and Cosmetic Act, Chapter II – definitions, Sec. 201(g)(1). Available at: http://www.fda.gov/opacom/laws/fdcact/fdcact1.htm (Accessed 18 February 2007).

2　US Food and Drug Administration, Center for Biologics Evaluation and Research, About CBER, http://www.fda.gov/cber/about.htm (Accessed 18 February 2007).

3　TITLE 21 Code of Federal Regulations, Food and Drugs, Chapter I – Food and Drug Administration, Department of Health and Human Services (continued), Part 600 Biological Products: General – Subpart A General Provisions, Sec. 600.3 Definitions.

4　US FDA Device Advice, http://www.fda.gov/cdrh/devadvice/312.html, Updated 6/30/1998 (Accessed 18 February 2007).

5　US Food and Drug Administration, Device Advice, http://www.fda.gov/cdrh/devadvice. Updated 4/23/03. (Accessed 18 February 2007).

6　Medical Device Directive – 93/42/EEC, http://www.mdc-ce.de/cert_md1.htm (Accessed 18 February 2007).

7　Annual Report FY 2005. III. Supporting Information. Review for Medical Devices and *In Vitro* Diagnostics Approval. (1) Approval Review for Medical Devices. Available at: http://www.pmda.go.jp/pdf/3title_supportinginfo.pdf (Accessed 19 February 2007).

8　TITLE 21 Code of Federal Regulations, Food and Drugs Chapter I Food and Drug Administration, Department of Health and Human Services, Subchapter A – General, Part 3 – Product Jurisdiction, Subpart A Assignment of Agency Component for Review of Premarket Applications, §3.2(e)(1-4) Definitions. http://www.fda.gov/oc/ombudsman/part3&5.htm#combo. (Accessed 27 February 2008).

9　EMEA Home Page, http://www.emea.eu. (Accessed 3 March 2008).

10　Globepharm, Links to Worldwide Regulatory Agencies, http://www.globepharm.org (Accessed 3 March 2008).

11　US Food and Drug Administration, FDA Organization, http://www.fda.gov/opacom/7org.html (Accessed 19 March 2007).

12　International Conference on Harmonisation of Technical Requirements for Registration of Pharmaceuticals for Human Use. About ICH, history and future, Available at: http://www.ich.org/cache/compo/276-254-1.html (Accessed 18 February 2007).

13　Global Harmonization Task Force, General Information, Summary Statement. Available at: http://www.ghtf.org/information/information.htm (Accessed 18 February 2007).

14　Guidance for Industry, Estimating the Maximum Safe Starting Dose in Initial Clinical Trials for Therapeutics in Adult Healthy Volunteers, Food and Drug Administration Center for Drug Evaluation and Research (CDER), July 2005, Pharmacology and Toxicology.

15　ICH Harmonised Tripartite Guideline E6(R1) Guideline for Good Clinical Practice: Consolidated Guidance, ICH June 1996, http://www/ich.org (Accessed 27 February 2007).

16　European Commission, Detailed guidance for the request for authorisation of a clinical trial on a medicinal product for human use to the competent authorities, notification of substantial amendments and declaration of the end of the trial, ENTR/F2/BL D(2003) CT1 Revision 2. October 2005.

17　Title 21 – Food and Drugs Chapter I – Food and Drug Administration Department of Health and Human Services Subchapter A – General, Part 50 Protection of Human Subjects, Subpart B Informed Consent of Human Subjects, Section 50.20 General requirements for informed consent. [46 FR 8951, Jan. 27, 1981, as amended at 64 FR 10942, Mar. 8, 1999].

18　Title 21 – Food and Drugs Chapter I – Food And Drug Administration Department of Health and Human Services Subchapter D – Drugs for Human Use Part 314 applications for FDA approval to market a new drug, Subpart B Applications, Section 314.50 Content and Format of an application, 314.50 (d)(5)(v). [50 FR 7493, Feb. 22, 1985].

19　International Conference on Harmonisation of Technical Requirements for Registration of Pharmaceuticals for Human Use, Efficacy – M4E(R1), Clinical overview and clinical summary of module 2 module 5: clinical study reports, Current Step 4 version dated 12 September 2002.

Getting started

Targeted Regulatory Writing Techniques. Clinical Documents for Drugs and Biologics,
edited by Linda Fossati Wood and MaryAnn Foote
© 2009 Birkhäuser Verlag Basel/Switzerland

Chapter 2
Regulatory writing tips

Linda Fossati Wood

MedWrite, Inc., Westford, Massachusetts, USA

Introduction

The discipline of regulatory writing is a combination of language, science or engineering, and law as applied to corporate objectives. General principles of good writing, particularly those pertaining to summarization of complex data, are certainly relevant to regulatory writing but are beyond the scope of this book. However, several strategies for focusing text are particularly helpful to the regulatory writer.

The goal of regulatory writing is to produce documents for submission to health authorities that are:

- Scientifically and editorially accurate
- Reflective of regulatory strategy and corporate goals
- In compliance with all applicable regulations and guidelines
- Clearly worded with respect to main messages

Several general rules of good writing and graphic design apply to regulatory writing for all submissions irrespective of submission type or region for intended submission. Understanding your intended audience and effectively using concepts such as visual logic, logical flow of content, and streamlining are essential to any well-written document. Management of global submissions is important and some suggestions are also included on this topic.

Audience characteristics

Consideration for your audience is of prime importance with any communication. The people who review regulated documents are generally scientists, clinicians, or engineers and most are highly educated. They are charged with protecting the safety of the general population, and most take their jobs and the attendant responsibilities seriously. You may, however, also assume they do not (understandably) have great familiarity with your product, may be rushed and overworked, may not speak English as a first language, and job training may have been brief. Therefore, the utmost in care must be taken to communicate simply and effectively. With this in mind, consider the following principles of regulatory writing as you write:

- Organization and navigation tools: Logical flow is essential to understanding information, and tools such as a table of contents are essential to finding information.
- Brevity: Focus on the key messages; too many details may actually obscure the point you are making.
- Conciseness: Summarize information to focus on relevant points.
- Clarity: Short, simple sentences convey complex messages well.
- Accuracy: Check the facts you are writing about.

General regulatory writing concepts

Visual logic

Many regulatory writing concepts are based on graphic design principles as applied to text and are intended to help your reader navigate through the text. A few of these principles are [1]:

Readers respond to a consistent page structure	Order, consistency, and simplicity constitute 'elegance' in design, as they reduce the reader's work. Templates establish design.
We search for "differences"	We have an evolutionary advantage in that we see differences in the environment. Bulleted or numbered lists, and bolded text exploit this advantage.
Visual stimulation keeps the reader awake	A pleasant appearance, coupled with visual interest, can draw the reader through the text and reduce the effort of reading. Use of tables, figures, and graphs provide visual interest and can help clarify a complex message.

Space attracts the eye   Good design requires that not all space be filled. White space can frame your message. Appropriate use of white space, a ragged right margin, and bulleted or numbered lists all help to make your messages stand out.

Side bar: Lessons learned

While the data are paramount to any regulatory submission, clean and clear writing is paramount to getting your message across. Regulatory writers will certainly meet resistance from some team members when they, the writers, begin to edit the submission. In our experience, few people will question statistical data, and most will admit to rudimentary knowledge of complex statistics; however, almost everyone fancies him or herself a writer. It is important for the writer to establish a good rapport with the rest of the team and to explain that if he/she finds the document difficult to read and comprehend, perhaps a regulatory reviewer will have similar problems.

The rule should be that the writer does not change concepts or conclusions without detailed discussion with the science-author but that editing for clarity and readability, formatting to eliminate brick walls, and formatting to increase visual logic are acceptable tasks for the regulatory writer. In our experience, initial resistance and skepticism can be replaced by eager anticipation of the writer's skills in this area if handled with sensitivity. Team members do not appreciate a 'schoolmarm' approach to their writing!

Logical flow and levels of detail

Conceptually, all submissions are pyramids, with a top-level summary at the apex (generally the cover letter, usually written by regulatory affairs), and text that gradually expands details as it moves toward the bottom of the pyramid. Therefore, the reader is first introduced to the product through a cover letter (that represents a high-level summary and therefore has the fewest details), and all subsequent text gradually expands on that summary information (all the way to the individual data points, which represent the greatest level of detail).

Document requirements for each level of this pyramid are defined individually by the three regions, by product category, by submission type, and even within companies, but a few general writing principles will help guide the writer. Each section of a submission and each document should begin with an introduction or executive summary (the top-level summary) that sets the stage for information to be discussed in the section and should include, as appropriate:

- Table of contents (unless the section is brief, for example less than 10 pages)
- Purpose of the section
- Product name
- Indication
- Brief description of information provided
- Applicable guidance documents/standards
- Summary of the conclusions drawn from the information presented
- Information common to all parts of the section to minimize repetition

Clinical information is also conceptually a pyramid. Figure 1 presents the conceptual summarization of clinical data, which starts with individual data points, moves into statistical data (in which the data points from each subject are summarized), into a

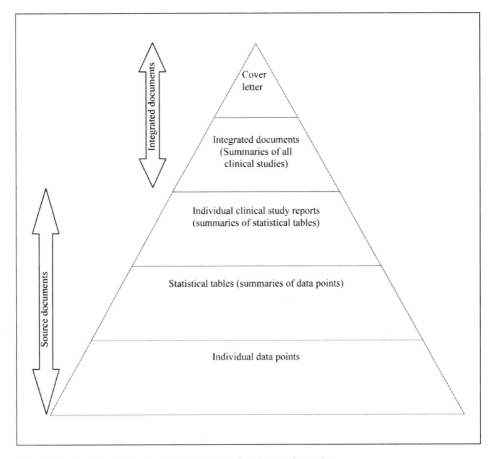

Figure 1. Logical flow and levels of summarization for clinical information

study report (which discusses the statistical data), then into integrated documents, which summarize information from several reports.

This concept of the development of clinical information is critical to keeping large masses of information straight and maintaining accuracy. It also helps to maintain writer sanity in the event that a team member requests the development of an integrated document before completion of a source document, an impossibility, given that information in integrated documents comes from the source documents.

Streamlining

Streamlining refers to the process of refining text to allow your key messages to stand out clearly. Removal of excess words, use of active voice, and simplicity and brevity of language all contribute to clarity of communication with your reader. Table 1 presents data as first written and then as edited for streamlining. Table 2 provides some examples of how to remove excess words to increase readability and clarity.

Table 1. Examples of text before and after editing for clarity, in the process of streamlining

---

Original text before streamlining:
As presented in Table 4b below, and consistent with the statistical analyses planned in the protocol and described in previous sections, the data were subjected to inferential testing to ascertain differences between treatment groups. Analytical methods (refer to Appendix 2 for details of the statistical analysis plan) failed to elicit discernible differences between the control group and the treatment group for any of the parameters selected and listed below.

Edited text after streamlining:
No statistically significant differences were noted between treatment groups (Table 4b).

Table 2. Less is usually better. Try to avoid nominalizations, complicated phrases, redundant expressions, jargon, and words such as "hereinafter, wheretofor, whatevertheheck"

As written	After editing
There is an urgent need to reevaluate….	We need to reevaluate….
…has an impact on…	…affects us…
…square in shape…	…square…
…red in color….	…red…
…8 hours time….	…8 hours…
…at this point in time…	…now…
…near future…	…soon…
…combine together…	…combine…
Past medical history…	Medical history…

Another method of streamlining is the use of bulleted lists [1]. Large blocks of text (also called brick walls) tend to visually bury information because all characters are given equal visual weight, and eventually the reader's eyes become fatigued and somnolence follows. Large blocks of print are particularly troubling for text containing numbers and units because they become difficult to read as the text wraps in the paragraph. Breaking text into discrete, visually demarcated lines can bring your most important points (key messages) into sharp focus. Table 3 illustrates the difference between large blocks of unbroken text and a bulleted list.

Table 3. Examples of brick-wall solid text and bulleted lists. The latter is easier to read and to grasp facts

Brick wall:
In single-dose studies, effects were largely noted at 600 and 1000 mg/kg (3600 and 6000 mg/m^2, respectively) in rats and included gastrointestinal changes, weight loss, and gross changes at necropsy. Ten days of repeated oral dosing in rats produced gastrointestinal-type clinical signs (125 mg/kg/day [750 mg/m^2/day]), body weight loss (100 mg/kg/day [600 mg/m^2/day] and 125 mg/kg/day), decreased food consumption (125 mg/kg/day), and gross necropsy findings (125 mg/kg/day).

Bulleted list:
Single dose effects in rats were the following:
• 600 and 1000 mg/kg (3600 and 6000 mg/m^2, respectively): gastrointestinal changes, weight loss, and gross changes at necropsy
Ten days of oral dosing in rats noted the following toxicities:
• 100 mg/kg/day (600 mg/m^2/day): body weight loss
• 125 mg/kg/day (750 mg/m^2/day): gastrointestinal-type clinical signs, body weight loss, decreased food consumption, and gross necropsy findings

Reference

1   White A, The Elements of Graphic Design. Space, Unity, Page Architecture, and Type. New York, NY. Allworth Press; 2002

Chapter 3.

Templates and style guides: The nuts and bolts of regulatory documents

James Yuen[1] and Debra L Rood[2]

[1]Moorpark, California, USA; [2]Liquent Certified Advanced Publisher, Simi Valley, California, USA

Introduction

The hallmarks of quality regulatory documents are clean data, a well-planned data analysis plan, clear writing, and a neat, well-organized document. Taken together, they lead to a high-quality document, may enhance or maintain the image of the sponsor, and can facilitate the review of documents by regulatory agencies. This chapter focuses on document appearance, describing best practices by which a high-quality, consistent, and professional appearance may be achieved using templates and style guides. It also focuses on the use of both paper files and electronic documents for regulatory submissions and on the creation and maintenance of templates and a style guide.

Document templates

Templates are a key component of a well-organized regulatory submission and fall into two categories:

- *Instructional templates:* Instructional templates provide the writer with instructions for writing sections of the document based on regulatory guidelines and the sponsor's interpretation of the regulatory guidelines. These templates are useful for documents for which there are regulatory guidance documents, such as protocols, investigator's brochures, clinical study reports, and informed consent forms. These templates also contain formatting styles and so apply consistent formatting to documents that will become part of a regulatory submission.
- *Property templates:* Property templates – also known as style templates – appear as blank pages. They do not contain instructions or guidance but instead contain formatting styles. The use of a property template applies consistent formatting to

documents that will become part of a regulatory submission, although the documents that use them may differ radically with respect to content. A property template is appropriate for the preparation of Investigational New Drug applications (IND); annual reports; Chemistry, Manufacturing, and Controls (CMC) sections; or Biologic License Applications or Supplements (BLA/BLS).

Template contents

The purpose of a template is to guide and direct the format and content of a regulatory document. A template should not be thought of as a fill-in-the-blank exercise. It may contain boilerplate text (ie, text that describes a concept that does not change and so therefore cannot be modified), and sections completely devoid of text. Neither implies a page of text with fill-in blanks. Each document, whether a protocol, investigator's brochure, clinical study report, informed consent form, or summary document must be considered carefully. The writer needs to consider the appropriate contents – and level of detail – for each section and subsection, even if it is to refer the reader to another section or to state that the expected test or procedure (for example) was not conducted as part of the study. The sponsor should have a policy on whether to permit the deletion of irrelevant sections or subsections as a best practice.

Each template should adhere to regulatory guidances, if applicable. That is, any topics specified in the guidance document should have a place in the template; however, with the exception of the guidance for the outline of the Common Technical Document (CTD) [1], the topics do not need to be in the same order as in the guidance document. See Appendices II, X, and XI for examples of protocol, clinical study report, and investigator's brochure outlines. Precise heading titles and the order of topics within the template may vary by sponsor and by product. Additionally, the depth of coverage of a topic is subject to interpretation by the sponsor. Thus, two sponsors can have different interpretations of a single section of a guidance document, or they might choose to implement selective portions of the guidance. The danger, of course, is in not providing information that regulatory agencies expect to find.

Guidance documents, while somewhat flexible, occasionally contain redundant sections. It is possible to avoid the redundancy by simply referring the reader (ie, regulatory reviewer) to another section of the document.

Clearly, the writer needs to read the boilerplate language and ensure that the text is accurate for the study. Boilerplate language should only be incorporated in templates when the language is not expected to change. It is recommended that any variance from approved boilerplate language in the template require a written explanation to the document review committee at the time of document review and approval by the committee.

One additional consideration during template development may facilitate the preparation of subsequent documents: Construct templates so that a single document can have multiple uses. For example, develop a format for the nonclinical toxicology section for the IND that allows that section to be simply cut and pasted into the investigator's brochure and later into Module 2 of the CTD. Also develop a template for individual document summaries that might be usable in documents that summarize the entire clinical or nonclinical program for an investigational product.

Instructional templates will be the most complex templates by virtue of the information provided. It is imperative that the template clearly distinguish between the various types of text in the template: for example, general instructions (found on the front page), additional information (bolded within brackets), boilerplate text (in regular typeface), and instructions (italicized text in double carets). The instructions are meant to be replaced by text in regular typeface. Figure 1 shows an example of a template with guidance text, boilerplate text, and instructions.

[Provide a statement indicating whether the study was performed in compliance with GCP (Good Clinical Practice). If not conducted in compliance with GCP, identify major deviations here.]

This study was conducted in accordance with <<specify applicable regulations>> and International Conference on Harmonisation (ICH) Good Clinical Practice (GCP) regulations/guidelines. Essential documents will be retained in accordance with ICH GCP.

1.X   Ethics

1.X.1 <<Independent Ethics Committee (IEC) or Institutional Review Board (IRB)>>

<<State that the study and all amendments were reviewed by an IEC or IRB at each study center. Provide a list of all IECs/IRBs consulted in Appendix X.>>

1.X.2 Ethical Conduct of Study

This study was conducted in accordance with <<specify applicable regulations>> and ICH GCP regulations/guidelines.

1.X.3 Subject Information and Consent

<<Describe how and when informed consent was obtained in relation to subject enrollment (eg, at allocation, prescreening).>>

Figure 1. Guidance text **[shown bracketed in bold]** should be deleted when writing the section. Boilerplate text must not be altered without an explanation to the document review committee and the committee's approval. Instructions <<shown in italics and enclosed in carats>> should be deleted during the writing process and replaced with the information requested; the replacement text should appear as regular type, not italic.

All templates should have a page of general instructions as the first page. Table 1 lists items that should appear on this general page. This page should contain a statement that indicates that the page should be removed when the writer prepares the document.

Table 1. General instructions to include with each template

Instruction	Protocol	Investi-gator's brochure	Clinical study report
The team should identify the lead author, who is responsible for ensuring completion of the document.	x	x	x
For phase 4 on-label studies, use the drug's trade name; otherwise use the investigational name.	x		x
For the primary and secondary endpoints, the report must be self-contained and must not cross-reference other documents.			x
It is not permissible to have one 1 subheading under a heading (ie, n.1 must have an accompanying n.2 at a minimum).	x	x	x
Italicized text surrounded by << >> is a description of information that should be provided by the author(s). The <<italicized>> text is to be deleted after pertinent information is incorporated in the document and should be replaced with regular text (ie, no italics; no bold). NOTE: Format for providing informational text should be determined by the sponsor.	x	x	x
Bracketed [bolded] text indicates special instructions to the author for that section or page and should be deleted in the final copy of the document. NOTE: Format for providing special instructions should be determined by the sponsor.	x	x	x
All other text is standard (ie, boilerplate) and should only be amended with documented justification. NOTE: To enforce this instruction, it is recommended that a memo to the document review committee be required.	x	x	x
Delete terms that are not applicable to the location (eg, international ethics committee or investigational review board).	x		x
Noncompliance with requirements described in a section will be discussed in that section.			x
Each major section (ie, 1., 2., 3.,) should start on a new page. NOTE: This is a sponsor's decision.		x	x
Each table and figure should have a reference line that cites the source of the information presented. The reference line must be specific enough to allow retrieval of the table or figure. Reference to other documents is permissible.		x	x
NOTE: General instructions should be provided as the first page of each template. The instructions must be customized for each template.			

Consistent formatting

Consistent formatting ties together a series of documents, giving them a similar look and feel. Years ago, before the advent of computers and sophisticated computer graphics programs, regulatory submissions commonly were assembled using documents and portions of documents from various sources. As might be expected, none of the pieces matched (ie, each department or group formatted its documents differently), and the result was a submission that appeared to come from a dozen different sponsors. The use of tape, whiteout, and photocopier to piece documents together certainly did not make for a consistent, attractive presentation. Now it is possible to provide submissions with a professional appearance with the use of consistent formatting by way of templates. Thus, the documents that originate from a wide variety of groups and departments – clinical development, nonclinical toxicology, commercial operations, and others – can look similar.

Use of templates will make documents consistent provided the appropriate styles are built into the template. Additionally, those writers who use the templates should receive training both on using styles and on formatting to ensure that they do not modify the styles in such a way as to create conflicts or confusion (eg, change font, spacing, or margins).

Create templates in house or purchase them from vendors?

Templates can either be created in house by a team of word-processing 'technical' experts, or they can be purchased from outside vendors. Experience has shown that, with technical expertise (ie, by someone who thoroughly understands the word-processing program and its intricacies), templates can be created and maintained in house. Otherwise, outside vendors with technical expertise can either create the templates or modify their templates to fit sponsor needs. Be forewarned, however, that subsequent revisions to the templates by the sponsor may necessitate a new round of major revisions to templates created by the vendor. The vendor may need to start anew to recreate templates that incorporate their latest guidance instructions with the sponsor's desired formatting and instructions.

Management review

Because document templates often cross departmental boundaries, it is imperative that the templates be reviewed by a committee comprising members of management, both to ensure the clarity of the template contents (ie, Does the template specify what is needed?) and to verify that any processes are workable (eg, Will the process of data collection be workable? Will the preparation of statistical tables affect the work of staff adversely?). Management needs to buy into the templates to effectively enforce their use in the future.

Availability of the templates

Once a template is approved by the document review committee, it needs to be made available for use by writers and other staff members. It is suggested that the templates be posted on the sponsor's internal Web site in a secured location. Writers should be able to download a fresh template when needed, but no one should be able to modify the posted template.

Users of the templates should be instructed to download a fresh template when starting to prepare new documents. Experience has shown that many users will work off old versions of templates, which often have changed quite drastically because of changes in regulations. The net result is a document that does not match the current template and, even worse, may contain text that pertains to a previous document (eg, a different protocol or product name). The best practice is to make the templates readily available to users on a secured Web site and to encourage users to download a fresh template for each new document.

Control of the templates

The chair of the document templates committee should be in control of the templates and should work with a highly skilled technical person to ensure that the templates are clear and contain no extraneous formatting or styles. As a matter of policy, the template committee chair should be the only one allowed to provide approved, revised templates to the sponsor's Webmaster, and conversely, the Webmaster should only accept revised templates from the template committee chair.

Style guide

All writers should have a style guide readily available. A style guide answers commonly asked questions about document style and format for all types of regulatory documents. As with the use of document templates, consistent use of a style guide can save the sponsor time and money, expedite the document development and review process, and help in producing quality regulatory submissions that enhance the sponsor's image and facilitate the product review process. Providing regulatory agencies with documents that are standard in format and style may facilitate the review by regulatory agencies and enhance the sponsor's reputation for submitting quality documentations [2]. This section will provide guidance on format and style considerations for customized style guides.

Standard style guides

A sponsor may choose either to create a company-specific style guide or to use a standard style guide. If the sponsor has a company-specific style guide for regulatory documents, that guide should be used by all staff in regulatory affairs, all writers, all contract writers, and anyone else who prepares documents that will be or may become part of a regulatory submission. A style guide should not be confused with either journal instructions to authors, which should be consulted before manuscripts are written, or corporate style guides, if any, that are used to ensure consistency in marketing materials. Even if a sponsor has a company-specific style guide, it is advisable to designate a standard style guide to which staff can refer for issues and questions not addressed in the sponsor's style guide.

Sponsors may not wish to produce and maintain a company-specific style guide. In such situations, the sponsor should settle on a widely used style guide as its standard, such as the *American Medical Association's Manual of Style* (2007) or *The Chicago Manual of Style* (2003). Table 2 lists other sources for standardization.

Table 2. Suggested style guides and standardization

- *AMA Manual of Style. A Guide for Authors and Editors*. 10th edn. New York: Oxford University Press; 2007.
- *The Chicago Manual of Style*. 15th edn. Chicago: University of Chicago Press; 2003.
- Davis NM (ed.) *Medical Abbreviations: 28,000 Conveniences at the Expense of Communications and Safety*. 13th edn. Huntingdon Valley, PA: Neil M Davis Associates; 2006.
- *Dorland's Illustrated Medical Dictionary*. 31st edn. Philadelphia: WB Saunders Co; 2007.
- *Merriam-Webster's Collegiate Dictionary*. 11th edn. Springfield MA: Merriam-Webster, Inc; 2003.
- *Stedman's Abbreviations, Acronyms & Symbols*. 3rd edn. Baltimore: Lippincott Williams & Wilkins; 2003.
- US National Library of Medicine. *List of Journals Indexed for Index Medicus 2008*. Available at: ftp://nlmpubs.nlm.nih.gov/online/journals/ljiweb.pdf. Accessed 27 February 2008.

Sponsor-specific topics in the style guide

The sponsor may choose to deviate from one of the standard style guides and may instead decide to incorporate some of its own preferences in a style guide. These sponsor preferences, along with key topics from any standard style guide, should comprise the sponsor's customized style guide. Table 3 lists suggested topics for a customized style guide irrespective of whether they conform to or differ from the sponsor's selected standard style guide reference. Sponsor page layout and style preferences in the style guide should be built into the document templates.

Table 3. Suggested topics for a customized style guide

Section	Subsection Topics

Document Layout
   • Components of a typical document
   • Page layout
	• Page margins
	• Document headers and footers

General Formatting
   • Text format
   • Use of US or UK English
   • Abbreviations and acronyms
   • Capitalization
   • Date format
   • Footnotes within text
	• Latin terms
	• Lists
	• Numbers
	• Superscripts and subscripts
	• Symbols
	• Units of measure

General Punctuation
   • Apostrophes (in eponyms)
   • Colons
   • Commas
   • Dashes
   • Hyphens
	• Parentheses and brackets
	• Periods
	• Quotation marks
	• Semicolons

Commonly Used Terms	
List of Preferred Terms and Correct Usage	
Company's Product Names • Capitalization of product names	• Hyphenation of product names

Tables
   • Placement
   • Borders and internal lines
   • Numbering
   • Column headings
   • Data format
	• Abbreviations in tables
	• Footnotes in tables
	• Source code
	• Multipage tables

Figures
   • Placement
   • Numbering
   • Titles
	• Figure legends
	• Source code

References
   • References citations in text
   • Reference lists
	• Copies of references

Attachments and Appendices
   • Citing in text
   • Order of appearance
	• Identification
	• Listing attachments and appendices

Guidance on document layout for paper and electronic submissions
Page layout
Regulatory documents are set up as a portrait page, commonly referred to as 'vertical page orientation.' The header is at the top of the page, and the footer is at the bottom. The lines of text run from the left to right margins. The landscape page, also known as 'horizontal page orientation,' may be used for tables and figures that are too wide to fit within the margins of a portrait page.

Page margins

Many sponsors submit marketing applications to regulatory agencies worldwide, so it is useful to prepare documents that will fit easily on both US letter paper (8.5"×11") and A4 paper (21.0×29.7 cm). Standard margin settings can be used to create a common text area of 9.25"×6" (23.5×15.24 cm) that will prevent text from reflowing when switching between paper sizes (Table 4).

Table 4. Page margins for paper and electronic submissions

	US letter paper (8.5" × 11" paper)				A4 paper (21.0 × 29.7 cm)			
	Portrait		Landscape		Portrait		Landscape	
Top	1.0"	2.54 cm	1.5"	3.81 cm	1.0"	2.54 cm	1.25"	3.17 cm
Bottom	0.75"	1.9 cm	1.0"	2.54 cm	1.44"	3.66 cm	1.0"	2.54 cm
Left	1.5"	3.81 cm	0.75"	1.9 cm	1.25"	3.17 cm	1.44"	3.66 cm
Right	1.0"	2.54 cm	1.0"	2.54 cm	1.0"	2.54 cm	1.0"	2.54 cm
Header[a]	0.25"	0.63 cm	0.75"	1.9 cm	0.5"	1.27 cm	0.75"	1.9 cm
Footer[a]	0.25"	0.63 cm	0.25"	0.63 cm	0.5"	1.27 cm	0.5"	1.27 cm
[a] This is a MS Word-specific setting. The measurement is from the edge of the page.								

Document headers and footers

Boilerplate headers and footers are often incorporated in each type of template and are specific to the document. These headers and footers often contain reference to the type of document (eg, clinical study report and name of sponsor). For regulatory submissions, it is often useful to have document headers and footers with identifying information that appears within the top and bottom margins of the page, respectively. The information contained in the headers and footers is uniform throughout the document, except for the item (or part) number and page number. Figures 2 and 3 show a sample header and a sample footer, respectively; however, the location of text within the headers and footers is subject to sponsor preference.

Product: Our First Drug
Clinical Study Report: 2007XXXX
Date: 21 April 2007   Page 1 of 1

Figure 2. Example of a header.

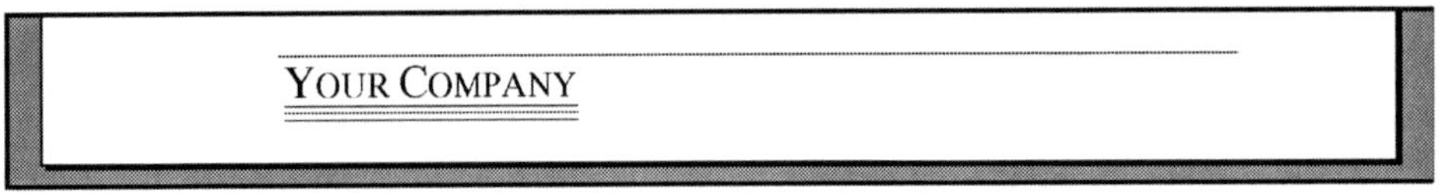

YOUR COMPANY

Figure 3. Example of a footer.

Page numbers
Hardcopy submissions
Documents created with templates often contain automatic page numbering, which can be in the format of 'x of y' (where 'x' is the page number and 'y' is the number of pages in the document). Documents are numbered from 1 beginning on the title page, using Arabic numerals. Avoid using Roman numerals to paginate the title page and table of contents; the use of Roman numerals complicates the report publishing process.

Electronic submissions
Per current guidance for providing electronic CTD (eCTD) submissions to the US FDA (ICH M4), every document should be numbered starting at page 1, except for individual literature references, where the existing journal page numbering is used. Arabic numerals should be used throughout.

There are two additional exceptions to the guidance. When a document is split because of its size (ie, >50 MB), the pagination of the subsequent file(s) should be continuous with that of the preceding file. When several small documents, each with its own pagination, are incorporated into a single file, it is unnecessary to renumber the documents into one page sequence, but each subdocument should have a bookmark to its starting page.

Guidance on general formatting
The text of submission documents should be black, Times New Roman, and 12 points, although a few other fonts are also acceptable [3]. Documents that are left justified (ie, block style, with paragraphs that are not indented) are easier to read and make the best use of space. Documents should use at least 1.5 line spacing, and paragraphs should contain extra spacing around each paragraph (eg, 6 points after each paragraph). These specifications should be built into the document templates' formatting styles.

Abbreviations, acronyms, and initialisms
Abbreviations – a shortened form of a written word, title, unit of measure, name, or compound term – are used to save space and to avoid cumbersome repetition of lengthy words and phrases. The use of abbreviations is strongly discouraged in clini-

cal documents. If needed, selective abbreviations should be used uniformly through-out an entire document, spelling out the abbreviations at first use (except very com-mon ones). Do not abbreviate pharmacokinetics or dosing intervals (eg, BID, QW) except in tables or when citing a dosing regimen in text (eg, The subject received drug XXX 5 mg BID).

Standard style guides are a good source of information on abbreviations, accept-able abbreviations, and punctuation of abbreviations. Generally eliminate periods in and after most abbreviations except when a period will clarify the meaning of the abbreviation (eg, 'No.' rather than 'No' for number).

Acronyms are a special type of abbreviation that comprises the initial letters of a term or parts of a term (eg, CTD, FDA, NDA). Initialisms are acronyms that can be pronounced like a word (eg, AIDS, NSAID). The general abbreviations guidelines and guidelines for punctuating abbreviations apply to acronyms and initialisms.

If a document contains many abbreviations and acronyms, list and define them at the beginning of the document, and define them again first use. A clinical study report synopsis is considered a stand-alone document; thus, abbreviations and acro-nyms must be defined both in the synopsis and in the body of the report.

Footnotes within text

A footnote is a note of reference, an explanation, or a comment placed at the bottom of the page (or table or figure) on which the referenced text appears. Footnotes are distinct from references, which cite a literature source and are provided as a list at the end of the document.

Numbers

Arabic numerals should be used for all numbers. For numbers less than 1, include a leading 0 (eg, 0.2%, $P<0.05$). Spell out cardinal or ordinal number at the begin-ning of a sentence, title, or header, ordinal numbers first to ninth (ie, first, not 1st or 1st), and adjacent numbers (eg, one 20-mg tablet). Avoid abbreviating units of time except as necessary in tables and figures, and do not place periods in abbreviations of units of measure unless the abbreviation might be confused with a word (eg, 'No.' for number). Units of measure are not pluralized (eg, 3 mg, not 3 mgs).

References

1 International Conference on Harmonisation (ICH). Guidance for Industry. M4. Organisation of the Common Technical Document for the Registration of Pharmaceuticals for Human Use. 13 January 2004. Available at http://www.fda.gov/cber/gdlns/m4ctd.pdf. Accessed 27 February 2008.
2 Foote MA, Hecker S, Style – what it is and how to get it. *DIA Forum* 2003; 39: 32–35.
3 US Department of Health and Human Services (US DHHS). Food and Drug Administration. Cen-ter for Drug Evaluation and Research (CDER) and Center for Biologics Evaluation and Research (CBER). Guidance for Industry: Electronic Format – General Considerations, Revision 1. October 2003. Available at http://www.fda.gov/cder/guidance/index.htm#electronic_submissions. Accessed 27 February 2008.

Targeted Regulatory Writing Techniques. Clinical Documents for Drugs and Biologics, edited by Linda Fossati Wood and MaryAnn Foote
© 2009 Birkhäuser Verlag Basel/Switzerland

Chapter 4.

Document review

Linda Fossati Wood[1] and MaryAnn Foote[2]

[1]Linda F. Wood, MedWrite, Inc., Westford, Massachusetts, USA; [2]MaryAnn Foote, MA Foote Associates, Westlake Village, California, USA

Introduction

Developing and writing a document for a regulatory submission is a group task. The writer may work on a given document alone for hours to days, but usually review meetings or round tables (ie, lengthy large group meetings) are scheduled on a regular basis to discuss and revise the writer's work. It is important for all team members to understand that the goal is production of a high-quality document that reflects corporate goals, regulatory strategy, and sound science.

Although the regulatory writer may be primarily responsible for writing text in a given document, it is not possible for one person to have responsibility for or knowledge of all the required scientific expertise, regulatory strategy, and corporate positioning necessary to write any regulated document. The process by which we bring others into the content we have written is review. Review allows other team members to provide input and to help shape the key messages so essential to product development. The review process may be broken down into several considerations:

- Team membership
- Time points for review
- Review mechanics
- Resolution of comments
- Sign-off of completed documents

Each team and each company approaches document preparation, review, and sign-off differently, but some key points can be made to help team efficiency.

Side bar: Lessons learned

Sign-off is often a major corporate problem. The problem may require a cultural shift from everyone knowing everything to one of information on a need-to-know basis. We are not suggesting tight silos, where no one from another group can look at documentation, because often it is necessary for other groups (eg, Marketing) to have some way to voice needs and concerns at various time points in the development of a product. Staff are fearful that if they do not have sign-off 'rights', they will know nothing and be out of the loop. A document system that allows all team members (nonclinical, clinical, statistics, regulatory, safety, marketing, etc) the chance to read and comment on a document in a finite time frame (ie, 24 hours to 10 days, depending upon many factors), but only three to five key people to sign off on the document often is the best solution to keeping everyone informed and the submission on track.

Team membership

The first step in review, before completion of the first draft of the document, is to establish who will be on the review team. The team will differ somewhat with every project, but for most clinical documents, certain functional areas are generally present. Team members should reflect those with great familiarity with the investigational product and with the project but also with the scientific, regulatory, or mathematical background and experience to meaningfully contribute.

The team must be trained in review techniques. Each team member must be perfectly clear on what is expected during his/her review of the document. Each team member should be reviewing content that is consistent with his/her educational background, job experience, and job title. It is not beneficial to have an entire team reviewing for verb tense and punctuation while leaving scientific or mathematical content to fend for itself. A statistical table has never checked itself; and no matter how meticulously we copy, paste, and label tables into a report, it is possible to mangle or mislabel a table. Table 1 presents review responsibilities for a clinical study report, which are relatively easy to assign based on functional area and job title. These responsibilities by necessity will change a bit for each study report, and for each team, but a general rule is that all content (methods, results, discussion, conclusions), and all types of content (medical, statistical, operational, and editorial) be assigned to at least one member of the team.

Table 1. Review responsibilities by job title

Job title	Review responsibilities for clinical study report
Medical writer	• Coordination of review, comment incorporation into document, resolution of conflicting comments • Editorial consistency of all sections – typographical errors, abbreviations, format, numbering, capitalization, agreement with template or style sheet, etc • Final report numbering, dates, page breaks, general appearance of document and volumes
Medical monitor	• Results, discussion, and conclusions sections • Clinical relevance of statistical results for efficacy • Relationship of drug to indication, reasons for results (bias, confounders, etc) • Interpretation of the relevance of serious adverse events and other safety data
Clinical research associates	• Methods, results, discussion, and conclusions sections • Accuracy of description of conduct of trial, version of protocol used, amendments included, patient narratives, Institutional Review Board (IRB) and investigator information, description of protocol deviations, withdrawals, adverse events
Statistician	• Methods sections only for statistical methods, results sections • Agreement of text with text tables and with statistical results and statistical tables • Reassurance that the correct version of the statistical tables was used for writing text and making text tables • Agreement of text table references to statistical tables and listings
Clinical project manager	• Methods sections, results sections for serious adverse event narratives • Accuracy of description of conduct of trial, version of protocol used, amendments included, patient narratives, batch numbers, IRB and investigator information
Safety officer	• Discussion and conclusions sections • Safety, risk/benefit information • Agreement between serious adverse event narratives and listings
Clinical phamacokineticist	• Methods, results, discussion, and conclusions sections • Accuracy of pharmacokinetics methods and results
Clinical immunologist	• Methods, results, discussion, and conclusions sections • Accuracy of immunology methods and results

Writers often use document-specific checklists to facilitate review and to ensure that all information is covered. Appendix I provides checklists for review of clinical protocols and clinical study reports.

Time points for review

Most documents undergo predictable stages: first draft, review and revision, second draft, review and revision, and a third and final draft. The general process for preparing regulatory documents should not allow for more than two drafts and a final version. Determining the target, the theme of this book, is critical for teams to quickly and efficiently reach the goal of quality submission documents. Experience suggests that, if a document requires more than two reviews/rewrites before the final version, planning was neglected, team members do not understand their functional area expertise roles, and timelines will not be met.

Review mechanics

Electronic distribution and use of 'Track Changes' and 'Comment' boxes have been useful for conveying review comments. Documents generally are distributed to all team members by e-mail with a date set for return of comments. Early drafts may require a week for proper review; final version should be allotted no more than 48 hours for comments.

Each team member is responsible for returning an electronic file with his/her comments. On occasion, a reviewer may not be comfortable with electronic conventions, or may not be in a position to review an electronic file easily. Hard copy review is fraught with problems for the writer, such as reading or interpreting the reviewer's comments, but may be necessary in these circumstances. All comments should be returned to the writer.

Resolution of comments

After receiving comments from all team members, the writer must consolidate comments and decide on a means by which conflicting comments, or issues that need additional discussion, should be resolved. The only thing more daunting than facing the monumental task of consolidating comments from perhaps 20 reviewers, is the thought of spending long hours in a windowless conference room agonizing over each comment on every page of the document. Although intuitively it is appealing to sit your team members down, lock the doors, and starting with page 1, reviewing, revising, and discussing every comment all the way through to the last page, this method of resolution should be avoided at all costs. In reality, even the most dedicated teams tend to deflate half way through and in a hurried manner rush to the end, having thoroughly dissected every comma and semicolon in the first half of the document but ignoring scientific content in the second half. This method is the most common method of document review despite the fact that it is acknowledged to be painful and generally ineffective.

Like many aspects of regulatory documentation, review is a marathon, not a sprint. Planning makes the difference. An alternative method of resolution is to di-

---

Side bar: Lessons learned

Document management processes and templates should be standardized across regions and changes suggested, discussed, and agreed to by all writing groups (and any other functional group charged with input, such as statistics). The concept of 'one document, many uses' can speed writing and reviewing time, and facilitate document management systems. Chapter 3 discusses standardized templates and boilerplate language.

---

vide comments into categories and present these in a PowerPoint presentation or Word file during the round table:

- Major comments that are resolved: a brief list so that reviewers can see that major points have been addressed
- Very focused list of conflicting comments summarized by type – section of the document, scientific, mathematical, or regulatory issues, etc

Two fundamental rules must be set at the beginning of a round table meeting, which are an attempt to maintain control of time and personalities and so accomplish the most important mission of review: to end up with a scientifically sound and accurate document. A statement should be made at the beginning of the meeting that all editorial and grammatical comments are under the purview of the writer and in the interest of time will not be discussed. In addition, all reviewer comments have been taken into consideration, although not all will appear on the lists presented. It may be necessary to restate these points several times during the review meeting to focus the team on content.

Sign-off procedures

Almost all companies require some sort of senior management sign-off for documents to be sent to regulatory agencies. Sign-off should signify that the person who is signing the document takes full responsibility for the accuracy of the data and the report. For this reason, many companies require the staff member who originally provided the data (ie, laboratory scientist, statistician) to sign the original report signifying responsibility for the accuracy of the data. In general, the writer is not responsible for the data provided for a given document, and any errors in transcription from original report should be identified and corrected in the internal auditing process. Sign-off should have grave legal implications (ie, if something goes wrong, signer does not have any excuses).

Sign-off can consume a significant amount of a time line. Corporate rules should be set that are followed for all documents of a regulatory submission. By the time a document requires senior management approval, all data problems, interpretation issues, and the like should have been resolved.

Another critical point is use of delegates: a message should be sent to all reviewers and all signers stating when the document will be available for review and comment, and when the document will be locked for further comment; also the message should state the day that sign-off will occur and that all signers must be in the office that day or must, without exception, provide the name of a delegate. The delegate's name is kept as a note to file, but sign-off occurs on the date prespecified. Signature 'approval' may be in the form of an e-mail if necessary (e-mail kept as note to file). While it is lovely to be a critical element in the drug development process, we live in uncertain times and no one person is or should be considered to have the sole authority to allow a document to move forward. Trusting in functional area expertise generally speeds the review process and engenders a feeling of collaboration with all departments.

Many writers instinctively recoil at the thought of document review, but this reaction can be avoided if teams adopt thoughtful, considerate approaches to the scientific review of documents and if teams allow writers to exercise their functional expertise in document preparation, language, style, and grammar.

Source documents

Chapter 5.

Protocols

Linda Fossati Wood

MedWrite, Inc., Westford, Massachusetts, USA

Introduction

The purpose of a protocol is to define the objectives, population, study design, procedures, outcome variables, and ethical conduct of product testing. Writing a protocol is a team effort, requiring cross-functional expertise, generally from clinical development, medical affairs, regulatory affairs, and statistics; health economics may be involved also. The process of writing a protocol involves a good deal of team interaction and negotiation to reach consensus. This chapter offers suggestions for the organization and content of protocols, while fully acknowledging that the team effort required to develop a protocol may result in a document that deviates from these suggestions. An example of a protocol outline is given in Appendix II.

The International Conference on Harmonisation (ICH) E6 Guideline [1] describes general content recommendations for a protocol to test drugs or biologics in humans. The description is brief with no attempt to direct the organization of the protocol. The lack of very specific guidance is an acknowledgment of the fact that a good protocol is the expression of scientific expertise and that this expression varies from study to study. The most important objective of a protocol is to communicate use of the investigational product to study-site personnel and to provide protection for study subjects. Therefore, simplicity and logical flow are paramount, and it is expected that these characteristics will vary from protocol to protocol.

Verb tense

Protocols are written in future tense ("serum sampling will be performed at baseline"), since they describe what will happen in the future. The only exception to this might be text in introductory paragraphs, which briefly describe results of nonclinical or clinical studies.

Title page, signature page, table of contents, list of abbreviations and definitions

While not part of the study design or conduct of the study, protocols have important auxiliary information that must be included in the protocol submitted to regulatory authorities.

Title page
An example of a protocol title page may be found in Appendix III. The title page of the protocol may contain several elements:

- Study title – A study title should state the stage of clinical development, the name of the investigational product, and the population to be studied. Study titles are descriptive and not promotional. Generic names are required with few exceptions (Table 1).
- Protocol identifying number – Procedures for applying numbering systems vary by company but often include codes for the name of the investigational product and other information for budgeting and organizing within the company.
- Names and addresses – It is important that investigators, study subjects, regulatory agencies, and institutional review/independent ethics committees know who is responsible for the conduct of the study.
- Name of the test drug – The generic name of the investigational product should be used, or the chemical name if the product does not have a generic name.
- Study design – The title page should briefly state the design of the study if this is not apparent from the title of the study. This chapter provides an in-depth description of study design characteristics. Some sponsors prefer to provide this information in the synopsis.
- Indication studied – An indication is a disease, syndrome, or diagnosis for which the product is intended, and the population for whom it is intended [2]. The indication statement reflects the culmination of all product testing (nonclinical and clinical) and the positioning of the product in the marketplace. Thus, it is probably the most important statement made about any product: it results from years of company research and negotiations with health authorities, and determines how marketing personnel are legally allowed to describe the product to customers. Because the indication statement is so vital to a product, it is generally the result of team consensus, the team comprising representatives from various departments such as medical and regulatory affairs, marketing, and business development. In addition, the indication statement is generally part of negotiations with health authorities during the process of submission review and product approval for marketing. Consequently, use of an indication statement requires that the writer work with the company's regulatory affairs department to determine the status and exact wording of the statement. Examples for the indication statement are provided in Table 2. Some sponsors prefer to provide this information in the synopsis.

- Dates – The title page requires the date of the protocol and the date of any protocol amendments. The date applied to the protocol should reflect the date of finalization of the original protocol. The original protocol is the first protocol for a specific study submitted to a health authority. Subsequent changes to the protocol are called amendments to the original protocol, and for the sake of regulatory compliance, it is very important to keep track of the original protocol version and the exact version of the amendments.

Table 1. Examples of protocol titles

- Phase 1, Single-center, Single-arm, Study of Panacea Acetate in Subjects with Progressive Alopecia
- Phase 3, Multicenter, Prospective, Randomized, Controlled Comparison of Panacea Acetate with Standard Therapy in Subjects with Progressive Alopecia

Table 2. Examples of indication statements

- XYZ, used in combination with infusional methotrexan, is indicated for the treatment of advanced carcinoma of the colon or rectum.
- ABC is indicated for reducing signs and symptoms of arthritis in subjects with moderate to severe rheumatoid arthritis; ABC can be taken alone or it can be used with methotrexan.

Signature page

A signature page is a convenient place to identify the sponsor's medical officer and indicate that the investigators charged with running the study have read the protocol. An example of a signature page is given in Appendix IV.

Table of contents

Every protocol should have a detailed and accurate table of contents. An inaccurate – or worse missing – table of contents is a disservice to your reader and may have ramifications for the ability to submit the document with a regulatory agency. Microsoft Word's Heading function allows for automatic generation of a table of contents.

For ease of navigation through a document, the table of contents headings should accurately reflect the content of the text under the heading. An example of a table of contents for a protocol is provided in Appendix II.

List of abbreviations and definitions

A list of abbreviations and definitions should be provided for the purpose of defining abbreviations. In the text of the protocol, spell out the word the first time it is used, and follow this with the abbreviation in parentheses. Each time this technique is done, the word and its abbreviation must be included in the list of abbreviations.

An example of a list of abbreviations is given in Appendix VI. Because sections are often read by different regulatory reviewers, it is good practice to treat each section of a submission as a stand-alone document and to treat the body of text as a new document, separate from the synopsis. Define all terms at first use in each section to assist the various reviewers.

Side bar: Lessons learned

It is always wise to avoid abbreviations when possible and to avoid all nonstandard abbreviations. The final protocol (or other part of the submission) should not read like alphabet soup. Although we often use abbreviations and jargon within our teams, we need to remember that the data and concepts are new and possibly unfamiliar to the reviewer. Do not make the reviewer guess what you mean or wonder whether the abbreviation ND means 'not determined' or 'no difference'.

Synopsis

The synopsis is a summary of all major characteristics of the protocol. An example is given in Appendix V. Because sections of the synopsis correspond to sections of a protocol, it is important that these sections be parallel in structure and information. Information concerning a specific section of a protocol is transferable to the corresponding section of the synopsis.

Although it seems intuitive that a synopsis should be written at the end, writing a protocol synopsis first is an effective way to outline fundamental characteristics of a study such as objectives, population, design, and outcome variables. Much of the administrative information found in a protocol (eg, study management, ethics, drug accountability, also called boilerplate information) is missing from a synopsis, resulting in a brief, easy-to-review document. It is usual for a team to develop a protocol synopsis first, discuss it at length, then use the information to complete the body of the protocol.

In general, the synopsis and the body of the protocol are treated like separate documents, in that words are spelled out the first time they are used in both, then the abbreviation added in parentheses.

Background

The background section of a protocol very briefly describes a few key characteristics of the investigational product for the convenience of study personnel. It is not intended to be an exhaustive review of all known data. In fact, extensive details are a detriment, since by the time the study is concluded, the information often is out of date and is, therefore, an inaccurate portrayal of the product. Greater detail may be found in the investigator's brochure (Chapter 7). References used should be cited at the end of the protocol in a reference section.

Background information comprises:

- Name and description of the investigational product(s) – The generic name, or the chemical name if the product does not have a generic name.
- A summary of nonclinical studies and clinical studies with potential significance to the study – a two- to three-page review of the nonclinical and clinical studies conducted to date on the investigational product, with emphasis on results that have significance for the personnel running the study. The text should refer the reader to the investigator's brochure for additional details.
- Summary of the known and potential risks and benefits, if any, to human subjects – a brief statement of any known or suspected safety issues and any potential benefits that may be associated with use of the investigational product. The text must clearly state the source of the information (another clinical study, nonclinical research, or literature reviews for similar products). All increased risks to which subjects might be exposed, and any methods by which these risks might be minimized, should be stated.
- Drug and intended study population – a rationale for the dose and route of administration and the duration of treatment based on a clearly stated source of information. A brief statement of the population characteristics, such as demographics (age and sex), disease characteristics, and the rationale for use in this population, should be stated briefly.
- Compliance statement – that the study will be conducted in compliance with the protocol, Good Clinical Practices (GCP), and the applicable regulatory requirement. A typical compliance sentence is "The study will be conducted in accordance with standards of Good Clinical Practice, as defined by the International Conference on Harmonisation and all applicable federal and local regulations."

Objectives

Clinical study objectives are a statement of the intended purpose of the study, the results of which should support the indication for use. Objectives tend to vary with the phase of development, as early development focuses on safety, and later development tends to focus on efficacy. Objectives are a statement of the overall purpose of the study and should be distinguished from outcome variables or endpoints, which are the means by which the objectives are measured.

Two fundamental objectives are tested in clinical studies: efficacy and safety (and tolerability). Efficacy is an evaluation of the product's ability to affect the disease or syndrome. Safety is an evaluation of whether or not the product may cause toxic or harmful effects. Safety is assessed in every clinical study, irrespective of developmental phase or product classification. Early-phase studies tend to assess primarily safety but may have an exploratory interest in efficacy. Later-phase studies assess efficacy in addition to safety. Additional objectives are feasibility (whether the product has any potential), pharmacokinetics (the effect of the body on a drug or biologic), pharmacodynamics (the effect of the drug or biologic on the body), or route of administration. A study may have several objectives, and these may be considered primary or secondary, based on the goals of the study. Examples of the objective statements and corresponding indication statements are provided in Table 3.

Table 3. Corresponding objective and indication statements

Objective	Indication
The primary objective of this phase 1 study is to assess the safety of panacea acetate when used in subjects with androgenetic alopecia. The secondary objective is to assess pharmacokinetic characteristics of the oral formulation.	For use in patients with androgenetic alopecia.
The primary objective of this phase 3 study is to assess the safety profile of ABC123. The secondary objective is to assess the ability of ABC123 to reduce the median fasting plasma triglycerides in subjects for whom diet has not worked.	For use in patients with lipoprotein lipase deficiency.

Methods

Study design
Every clinical study has a design, a system by which the methods of testing and the study subjects on whom the product will be tested is constructed. The scientific integrity of the study and the credibility of the resulting data depend substantially on the study design. All study design features intend to control anything that will attenuate the ability to compare groups. The study design considered to be the gold standard for industry research is the prospective, randomized, well-controlled design, often referred to as a randomized controlled trial or RCT. Many study designs are possible, but the regulatory writer need only be familiar with a few basics of study design most commonly used to assess drugs and biologics.

Phases of study
Phases of a study for drugs and biologics are discussed in Chapter 1 and are divided into premarketing phases (phase 1, phase 2, and phase 3) and a postmarketing phase (phase 4). The phase of the study influences many design features, such as the number of subjects and the frequency and type of measurements.

Enrollment

Study design is described in terms of enrollment of the study groups in relation to time (prospective and in parallel being the optimal situation), the number of investigational centers to be included, the use of a control group(s), the method by which bias in the data is minimized (randomization and blinding), the type of study subject to be enrolled, and how the objectives of the study will be assessed (outcome variables).

Study enrollment can be either prospective or retrospective. Prospective enrollment refers to the collection of the data going forward in time. Retrospective studies analyze data that have been collected before the study started (ie, as part of another study or demographic data) and are studied under retrospective designs. These study designs include case control (subjects with a disease compared with subjects with similar characteristics but without the disease) or historical control (subjects in a prospectively enrolled group are compared with subjects selected from existing documentation, generally medical records or literature). Enrollment may also be described as parallel, in which enrollment of the study groups occurs at the same time, in contrast to enrollment of one group after another.

Controls

A controlled study has two or more comparator groups and compares the results of treatment with the investigational drug with the results of treatment with a comparator treatment. Four types of controls have been defined: no treatment, placebo control (no active ingredient), active control treatment (another product), and historical control (data from previous studies or from literature).

Cohorts

Study cohorts, sometimes referred to as study groups, are another important part of the study design. In a single-cohort study, only one investigational product is used. When two or more cohorts are used, the investigational product is compared with one or more other products or no product. In oncology studies, dosing cohorts also may be used to compare different doses. Dosing cohorts are often used in oncology studies in which three subjects are administered investigational drug (cohort 1), and if the investigational drug does not have severe side effects, cohort 2 (a second set of three subjects) will receive a higher dose of the drug. Subsequent cohorts continue to receive higher doses (dose escalation) until the maximum tolerated dose (MTD, the highest dose considered to be safe) is reached.

Randomization

Randomization is the process that assigns subjects by chance, rather than by choice, to either the investigational product group or the control group [3]. Randomization is an attempt to produce comparable groups by distributing subject characteristics (such as sex, age, and severity of disease) evenly across the study groups. The goal of balancing these characteristics across groups is to reduce the potential for study results that are influenced by differences in subject groups and therefore minimize bias in the data. Randomization is generally performed using a computer-generated randomization scheme. Subjects are allocated to a cohort without their knowledge of the treatment to be given and without the study personnel's knowledge of the treatment the subjects will receive.

Blinding/masking

Blinding or masking refers to intentional labeling of the investigational drug, or to study design characteristics that hide the identity of the product administered to study subjects with the intent to minimize or avoid bias in the data. The word 'masking' is preferred in ophthalmic studies, except in the European Union, in which the word 'blinding' is used irrespective of the product under investigation. Blinding can occur on several levels. In a single-blind study, either the subject or the investigator is blinded to treatment; in double-blind studies, both the subject and the investigator are blinded to treatment [3]. Third-party blinding is used when blinding of the subject and the investigator is not possible and a third party responsible for observing study results, such as laboratory values or radio-imaging scans, is blinded. A third-party blind design is particularly helpful if unintentional removal of blinding is possible. This occurs when a clinical sign (such as a rash, a decrease in blood pressure) would allow either the subject or the investigator to guess the identity of the treatment. Knowledge of the treatment administered may potentially bias study results.

The protocol should describe procedures for maintenance of study treatment randomization codes and procedures for breaking codes, necessary in the advent of an adverse event.

Number of centers

Clinical trials are conducted by practicing physicians in hospitals, medical centers, clinics, and medical offices. Studies may be either single center (one investigational site) or multicenter (more than one investigational site). Early studies in drugs and biologics (phase 1) are often conducted at a single center that specializes in these types of studies (ie, a phase 1 house). As product development progresses and the sample size of the study increases, multicenter studies are common as more than one site is required to find sufficient numbers of subjects. Very often a study has centers across geographical regions.

Study population selection and withdrawal

Each study defines the characteristics of subjects in whom the product will be tested. A specific patient profile is necessary to test the product in subjects who are most likely to show a beneficial effect but who are also unlikely to be harmed by the investigational product. These definitions are described in inclusion and exclusion criteria and differ markedly by both the product and the disease state being studied.

The wording of inclusion and exclusion criteria may either be so restrictive (ie, tight) that few subjects are eligible for enrollment or so loose (ie, open) that too much data variability exists to enable meaningful statistical analyses. Tight criteria can slow study enrollment and open criteria can speed up enrollment but at the expense of meaningful information. Inclusion and exclusion criteria are generally numbered, not bulleted, which makes it easier to report deviations to criteria in the final clinical study report or changes to criteria in an amendment.

Inclusion criteria are generally worded as something the subjects must have or must be. Exclusion criteria are generally worded as something the subjects must not have, or must not be (Table 4). Although it may be tempting to place the same information in both inclusion and exclusion criteria (ie, subjects must be at least 18 years old in the inclusion criterion, and must not be under 18 years old in the exclusion criterion), eventually a revision will be made to the protocol and only one of these criterion will be changed, leading to a discrepancy. Discrepancies may diminish the ability of the clinical study to collect meaningful data.

Table 4. Examples of inclusion and exclusion criteria

Inclusion criteria	Exclusion criteria
Eligible subjects must meet the following criteria to be enrolled in the study: 1. Aged ≥18 years 2. Histologic or cytologic diagnosis of advanced cancer 3. Adequate renal function (defined as BUN <2 times upper limit of normal) 4. Adequate hepatic function (defined as AST/ALT <2 times upper limit of normal) 5. Eastern Cooperative Oncology Group (ECOG) performance status 0–1 6. Life expectancy of at least 3 months 7. Signed and dated written informed consent	Eligible subjects must not have any of the following to be enrolled in the study: 1. Active infections or serious intercurrent illness, including hepatitis B or C 2. Presence of unstable angina, recent myocardial infarction (within the previous 6 months) or use of ongoing maintenance therapy for life-threatening arrhythmia 3. Known hypersensitivity to this class of drugs 4. Pregnant or breastfeeding women, women who are of childbearing potential, and women who are not using an effective method of either barrier or hormonal contraceptives 5. Any issue that, in the opinion of the investigator, would render the subject unsuitable for study participation

Stopping rules

A description of the stopping rules or discontinuation criteria for individual subjects, parts of the study, and the entire study should be included in the protocol. Subjects are always allowed to discontinue from a study at any point in time if they wish to, without need to supply an explanation, and without repercussions. The sponsor and the investigator may also stop the study or withdraw a subject for a variety of reasons. Reasons for stopping a trial range from safety concerns to financial difficulties for the company.

Outcome variables: Safety and efficacy

Outcome variables, also called endpoints, are the measurements used to determine whether the study has met its objectives. Therefore, if the study objectives state that safety and efficacy are being assessed, then the outcome variables should include a list of safety endpoints and a list of efficacy endpoints. Examples of presentation of study objectives (the purpose of the study) with the corresponding outcome variables (the means by which the purpose of the study is assessed) are presented in Table 5.

Table 5. Corresponding objective and outcome variable

Objective	Corresponding outcome variable
Efficacy • To determine the efficacy of daily subcutaneous XYZ when administered in combination with DDF in women with ovarian cancer whose disease has progressed or recurred	*Efficacy* • Progression-free survival using Response Evaluation Criteria in Solid Tumors (RECIST) criteria • Objective response rate (for those women who have clinically evident disease as defined by a lesion of ≥2 cm by CT or MRI, or ≥1 cm on spiral CT) • Duration of objective response • Changes in serum CA555 concentrations • Changes in quality-of-life measurement
Safety • To determine the safety of XYZ in combination with DDF under the conditions of the study	*Safety* • Vital signs and weight • Symptom-directed physical examinations • Adverse event reporting • Clinical laboratory testing: liver and renal function, urinalysis, hematology • Left ventricular ejection fraction

Study schedule

All clinical studies have several stages during which specified procedures are performed. A study generally starts by screening subjects to determine eligibility for enrollment (using inclusion and exclusion criteria). Baseline examinations for eligible subjects are conducted to establish each study subject's disease status before administration of the investigational product. The protocol defines dosing periods; times for measurement of outcome variables; and a follow-up period, the time after dosing with the investigational product has ended, but during which the subject is still under observation. A visual representation of study periods (called a study schema) may be found in Figure 1.

Assessment of outcome variables after baseline is performed at specified time points to determine how baseline data have changed. The text of the protocol describes exactly which assessments are to be measured for each time point, and the description is augmented by a table (called by various names such as a study schedule or flow chart) displaying the time points and all assessments. An example of a

Side bar: Lessons learned

While most journals require, as condition to publish, that a phase 2, 3, or 4 clinical study be registered on a publicly accessible database, such as www.clinicaltrials.gov, the Food and Drug Administration (FDA) now mandates registration. Public Law 110-85, Section 810 expands the role of clinicaltrials.gov, expands the scope of clinical trials that must be registered, increases the number of registration fields that must be submitted, and sets penalties for noncompliance. Discussion of clinical trial registries is beyond the scope of this book; however, it is now necessary to have a process in place for registering clinical trials at the time they are written.

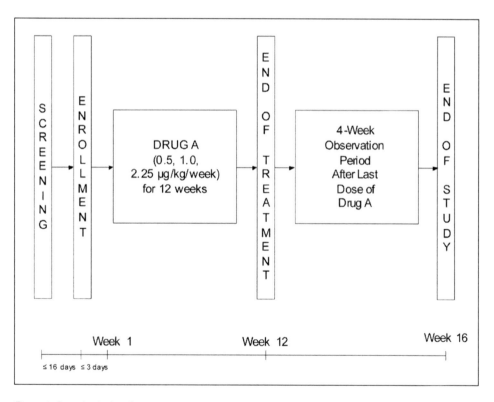

Figure 1. Sample study schema

study schedule is presented in Table 6. Evaluations to be done are listed in the first column of the table, and the visit when the evaluation will be done is marked with an X. Study schedules are used by investigational site personnel as a snapshot of study design and to determine resource allocation at the study site; thus, they are important to include in a protocol.

Table 6. Example of study schedule or flow chart

	Before treatment		21-day cycles									End of study
			Days cycle 1					Days cycles 2–4				
	14 days	72 hours	1	2	8	15	21	1	8	15	21	
Xerimax			X					X				
Informed consent	X											
Medical history	X	X										
Physical exam		X	X	X	X			X	X			X
Height		X										
Weight		X	X	X	X			X	X			X
Vital signs		X	X	X	X	X		X	X	X		X
ECG	X		X					X				X
ECOG PS	X		X		X			X				X
Labs	X		X		X	X		X	X	X		X
Serum pregnancy test	X											X
Concomitant medications	X	X	X	X	X	X	X	X	X	X	X	X
Pharmaco-kinetics			X					X				
Tumor imaging										X		X
Adverse events			X	X	X	X	X	X	X	X	X	X

Treatment administration

The investigational product should be described in terms of the quantity, route, and timing of administration. The information should include the dose (including units), dosage form (eg, capsule, tablet, injectable, ointment, patch), route of administration (eg, oral, intravenous, subcutaneous, topical, sublingual), and frequency of administration (eg, daily, weekly, 3 times weekly, once monthly), ingredients (active and inactive), and instructions for preparation, administration, storage, and handling. In addition, accountability (the process by which investigational supplies will be counted, tracked, and disposed of) should be described.

Study duration

All studies have a defined duration for each subject and an estimated duration for the entire study. Duration by subject is defined as the period from first screening visit or baseline procedure until the end of all assessments, including follow-up for adverse events. Although each subject's experience may vary, it is relatively easy to define subject duration on study by adding up the number of days/weeks/years on the study schedule. For example, if screening and baseline procedures are to start 2 weeks before initiation of dosing, treatment lasts for 3 months, and follow-up procedures are to be 2 weeks after the last dose, subject duration would be approximately 16 weeks (or 4 months).

Duration for a study is defined as the first day of the first subject's assessment as recorded on a case report form, through the last subject's last observation as recorded on the case report form. Because study duration must be dependent on the rate of subject enrollment as well as each subject's time on the study, it is more difficult to estimate than a single subject's duration on study. In a study of 30 subjects who are expected to be enrolled at the rate of five subjects per month and who will have an estimated duration on study of 16 weeks, the estimated duration of the study would be 10 months. This simple mathematical calculation seems appealing but would generally be considered extremely misleading, because of a number of factors that slow enrollment (eg, investigators who do not have the study population promised; subject and site personnel vacations; changes in staff, equipment, or the physical facility). In addition, subjects occasionally experience adverse events that require extended follow-up, or they miss scheduled visits and must be seen at a later date. Therefore, estimates of study duration always include additional time. The amount of additional time is based on the sponsor's experience with particular study sites, disease, general availability of subjects who fit the inclusion/exclusion criteria, and time to complete administrative details, such as contracts and Institutional Review Board (IRB) approvals.

In a perfect world, studies would be designed such that all subjects would enroll quickly, study periods would have a defined and short-term end because subjects would be completely cured of disease, and the investigational product would be ready for a marketing submission within the space of a few months to a year. Clearly, this is not the case, so protocols tend to have a number of considerations that obscure a quick and easy end to the study.

Requirements for long-term follow-up for outcomes such as survival may be required based on negotiations with a health authority or the sponsor's needs for labeling but can extend study duration up to several years, thereby delaying a regulatory submission. It may be possible to define study periods in a protocol such that duration is divided into two segments: the first, a relatively short time period, and the second, a longer time period. Data from the first segment could be fully analyzed and reported in a clinical study report quickly and used in the regulatory submission. Data from the second segment could be analyzed and reported later (ie, 5 or more years in some survival studies of large study populations). In this scenario, protocols can be written as one protocol with two segments or parts (ie, part A and part B, short and extended, respectively); or as two protocols (short and extended) but all subjects from the first protocol are to be enrolled in the second protocol (ie, a rollover study protocol).

Study monitoring

Sponsors are required by law to monitor all clinical trials [4]. Monitoring consists of a set of procedures that include visiting the investigational site before enrollment of the first subject, assessing whether personnel and the physical facility are adequate to conduct the study, managing investigational supplies, and making periodic visits to review study data and general trial conduct. Monitors collaborate with site personnel to ensure safe and compliant conduct of the study.

Descriptions of monitoring should be very general in the protocol because good monitoring procedures vary widely and cannot be rigidly defined. It is sufficient to say that monitoring will be performed, and cite the guidance used. The clinical operations staff or regulatory affairs should be able to provide information concerning which clinical trial guidance will be applicable, as this varies by region.

Data management

The data management section of the protocol should be written by data management personnel. The regulatory writer usually has a limited role in this section, other than that of ensuring editorial consistency. The section should be brief (one to two paragraphs): detailed descriptions are best left to operational manuals.

Statistical analysis

The statistical analysis section should be written by a statistician. The role of the regulatory writer is to review the section to ascertain whether or not a description of analysis of all outcome variables has been included.

Ethics

The purpose of the ethics section of the protocol is to describe compliance with local and international laws during the conduct of human research. Three statements are required, but of course are generally expanded to better meet the specific requirements of the study:

- The study and any amendments will be reviewed by an Independent Ethics Committee (IEC, the name used in Europe) or Institutional Review Board (IRB, the name used in Japan and the United States)
- The study will be conducted in accordance with the ethical principles that have their origins in the Declaration of Helsinki
- Informed consent will be obtained for subjects enrolled in the study

Study administrative procedures

A protocol provides basic procedural details for record retention and confidentiality, financing and insurance (unless this information is covered by another document), and publication policy. Many of these details, however, are generally found in other clinical study-associated documents such as the contractual agreement with the investigator. Descriptions in these sections may be brief and will vary greatly with the company.

Amendments to the protocol

A revision to the protocol that might significantly affect subject safety, the scope of the investigation, or the scientific quality of the study cannot be implemented until it is written as an amendment to the protocol and submitted to health authorities [5].

A protocol amendment usually requires two documents: an amendment (which describes the reason for the revisions and a detailed account of all revisions), and a revised protocol (with all revisions from the amendment included and the amendment date on the title page. Amendments must be written in such as way as to allow the reader to see the wording of the protocol before and after the revision. A common format has new text presented in **bold** typeface and deleted text presented with a ~~Strikethrough~~. A sample amendment is given in Appendix VII.

An amendment should have information that makes it easily identifiable and associated with a specific protocol, and the purpose should be clear. A suggested approach includes use of a title page with protocol title and protocol identifier, and a signature page with lines for both the sponsor's medical officer and the investigator(s).

Side bar: Lessons learned

A good protocol not only is scientifically sound, facilitates a well-run clinical trial, and sets the stage for meaningful statistical analyses, it also lends itself easily to development of the methods sections of a clinical study report (Chapter 6). The content of the clinical protocol forms the basis for Sections 1–9 in the ICH E3 guideline, so while writing the protocol, the writer should use the ICH E3 guideline to ensure that content areas are covered. This protocol content is most easily turned into Sections 1–9 of the clinical study report if the protocol is organized similarly to the guideline. But logical flow for the operational functions of a clinical study is not necessarily the same as logical flow for reporting of study data. If site personnel are confused, the study is run poorly, and trial data are not collected properly; no amount of effort put into writing the clinical study report will be able to save the situation. Based on many attempts to organize a clinical protocol to fit easily into the ICH E3 clinical study report outline for Sections 1–9, we suggest that the writer work closely with clinical operations personnel to ascertain the best way to organize the protocol so that it can be easily read by the site personnel and leave the difficulties inherent in writing the clinical study report until later. Site personnel tend to refer to the study schedule (also called the flow chart) and the inclusion/exclusion criteria more than most other sections of a protocol, so these should be easy to locate. Numbering of the inclusion/exclusion criteria (as opposed to using bullets) is used to provide an easy way to identify the different criteria.

References

1   ICH E6, Guidance for Industry, Good Clinical Practice: Consolidated Guidance, April 1996. http://www.fda.gov/cder/guidance/959fnl.pdf. (Accessed 5 March 2008).
2   Title 21 – Food and Drugs Chapter I – Food and Drug Administration Department of Health and Human Services Subpart B – Labeling Requirements for Prescription Drugs and/or Insulin. Sec. 201.56 Requirements on content and format of labeling for human prescription drug and biological products, April 1, 2007. http://www.accessdata.fda.gov/scripts/cdrh/cfdocs/cfcfr/CFRSearch.cfm?fr=201.56 (Accessed 9 March 2008.)
3   National Cancer Institute, US National Institutes of Health. What is randomization? www.cancer.gov. (Accessed 5 March 2008).
4   Title 21 – Food and Drugs Chapter I – Food and Drug Administration Department of Health and Human Services Subpart D – Drugs Responsibilities of Sponsors and Investigators. §312.50 General responsibilities of sponsors, April 1, 2007. http://www.accessdata.fda.gov/scripts/cdrh/cfdocs/cfCFR/CFRSearch.cfm?CFRPart=312&showFR=1&subpartNode=21:5.0.1.1.3.4. (Accessed 9 March 2008).
5   Title 21 – Food and Drugs Chapter I – Food and Drug Administration Department of Health and Human Services. Subpart D – Drugs for Human Use. §312.30 Protocol Amendments. http://www.accessdata.fda.gov/scripts/cdrh/cfdocs/cfcfr/CFRSearch.cfm?fr=312.30. (Accessed 9 March 2008).

Targeted Regulatory Writing Techniques. Clinical Documents for Drugs and Biologics, edited by Linda Fossati Wood and MaryAnn Foote
© 2009 Birkhäuser Verlag Basel/Switzerland

Chapter 6.

Clinical study reports

Linda Fossati Wood

MedWrite, Inc., Westford, Massachusetts, USA

Introduction

The clinical study report describes the results of a single human study and thus represents the most fundamental building block in a drug product's argument for use in humans. The results of all human trials conducted by a drug or biologics company must be recorded in some type of report, although the type of reporting may vary. This chapter describes a clinical study report as defined by the International Conference on Harmonisation (ICH) in the E3 Guideline [1], which is the type of report used for most clinical studies conducted in the three major geographic regions (Europe, Japan, and the United States) and is referred to as a "full" clinical study report in this chapter.

This chapter also describes two alternatives to the full clinical study report. The Food and Drug Administration (FDA) in the United States has provisions for two additional report types: an abbreviated report, and a synopsis [2]. Table 1 presents a description of each of these report types. A full ICH E3 clinical study report is a labor- and time-intensive task and is intended to fully support both safety and effectiveness for product labeling; however, many product development programs include clinical studies that for several reasons (eg, inadequate enrollment, poor study design, bias in the data resulting from lack of control of confounders, or noncompliance with Good Clinical Practices) do not meaningfully contribute. The purpose of FDA's provisions for abbreviated reports and synopses is to reduce needless work on a full report.

The decision to submit an abbreviated report or a synopsis instead of a full report is generally the responsibility of regulatory affairs, and is often the result of negotiations with FDA.

Table 1. Clinical study report types for submission in the United States

Report type	Description
Full study reports	Complete ICH E3 reports are submitted for all clinical and human pharmacology investigations that contribute to the evaluation of effectiveness for the proposed indication, or that otherwise support information included in labeling.
Abbreviated reports	Submitted for studies that are not intended to contribute to the evaluation of product effectiveness or provide definitive information on clinical pharmacology, but about which the reviewer needs sufficient information to determine that the study results do not, in fact, cast doubt on the effectiveness claims or the description of the clinical pharmacology. Abbreviated reports should contain all the safety information included in a full report.
Synopses	Submitted for studies that are not relevant to evaluation of product effectiveness or clinical pharmacology, but that provide information the reviewer needs to evaluate the safety data from the study. Complete safety information from a study submitted in synopsis format should be included in the Integrated Summary of Safety (ISS) [21 CFR 314.50(d)(5)(vi)(a)] for drug products and for biologic products where an ISS is included in the application. For biologic product applications not containing an ISS, the safety information for studies submitted in synopsis format should be appended to the synopsis.

Full clinical study reports

Full clinical study reports should be written using the ICH E3 Guideline for Industry, Structure and Content of Clinical Study Reports [1] and in close collaboration with a multidisciplinary team that minimally includes medical, statistical, and regulatory expertise. Careful and thoughtful interpretation of statistical data, within the framework of clinical medicine, and with the objective of supporting the drug's indication statement, is the goal for a study report.

The ICH E3 Guideline provides solid advice for content, but contains a table of contents that has been misconstrued as a directive from the ICH Working Group as an outline that must be followed exactly, with potential for disaster for anyone who dares to deviate. The rumors suggesting such rigidity in this guideline have caused a fair amount of angst in the industry, as efforts to force every product into the same outline have resulted in some rather obscure pairings of headings and content. These odd combinations of heading and content only serve to confuse health authority reviewers as they attempt to locate information.

The working group for this guideline never intended this interpretation of the table of contents and, in fact, vigorously defends the right to modify the outline to best characterize a product. Adherence to the general organization as suggested in the first and second headings has been found to be useful, however, as reference to these numbers is becoming easily recognizable in the industry. This chapter describes writing a clinical study report in terms of these first- and second-level section headings, but with the caveat that organization of the report should be dictated by logic and good communication principles.

Good communication principles in scientific writing have previously been characterized by a small group of medical journal editors who met informally in Vancouver, British Columbia, in 1978. During this meeting, the group (which later became known as the Vancouver Group) established guidelines for the logical flow of manuscripts submitted to their journals. Although this style of organization was originally applied to published articles, it has since become the standard for all research written in industry.

The logical flow as defined by this group is an introduction, methods section, results description, and discussion that is referred to as the 'IMRaD' structure and is considered a direct reflection of the process of scientific discovery [3]. This flow of logical thought is understood by scientists worldwide and should be used whenever possible when reporting research results. All study report outlines developed globally by regulatory agencies use this fundamental organization of content as a baseline for logical communication. The ICH E3 outline roughly corresponds to the IMRaD structure as shown in Table 2.

Table 2. Corresponding sections of IMRaD style for journal articles and ICH E3 sections

IMRaD	ICH E3 section
Introduction	7 Introduction
Methods	5 Ethics 6 Investigators and study administrative structure 8 Study objectives 9 Investigational plan
Results	10 Study subjects 11 Efficacy evaluation 12 Safety evaluation
Discussion	13 Discussion and overall conclusions

Therefore, if one follows the first-level headings in the outline supplied in ICH E3 for human studies, the reports will be consistent with the fundamentals of the IMRaD structure. Subsections of the ICH E3 outline may be modified as appropriate for a particular product and study, in the interest of clear communication.

Verb tense

In general, clinical study reports are written using past tense ("blood pressure was measured every 4 hours") as they report events that have occurred in the past. Exceptions to this in the sections corresponding to ICH E3 Section 9 of the report are:

- Reports of studies that are ongoing: If the report covers a study that is not completed (ie, all study subjects have not had the last possible data point collected or the study has not been officially terminated), then some of the text may reflect this.
- Protocol deviations: Conduct of the study, as described in the original protocol, can change in one of two ways: from a legitimate change described in a protocol amendment (Chapter 5) or from a deviation. Deviations are changes from the letter of the protocol in some aspect of study conduct and range from minor mistakes that have little if any impact on subject safety or the validity of study results to major changes that endanger subjects, invalidate results, or represent intention to commit fraud. Due to the complexity of studying humans, protocol deviations occur frequently and require a bit of planning on the part of the regulatory writer.
- Changes in verb tense due to deviations are complicated, but may be resolved using one of two solutions:
  - Past tense ("blood samples were taken at baseline"). Although use of past tense is intuitively correct, it requires that the writer obtain a list of every deviation then modify all text to reflect all instances, focusing primarily on deviations that might affect outcomes (major deviations). It also requires that the list of deviations be accurate and complete and that deviations considered to be major be identified and separated from most of the deviations that tend to be minor and of little consequence. Changes in verb tense to accommodate all minor deviations in even the most perfectly run clinical trial would require substantial time and resources, and likely for little benefit. Study reports are subject to audit by health authorities, and inaccuracies do not benefit product approval.
  - Past intention, which may be used throughout all methods sections of the clinical study report ("pharmacokinetic sampling was to be performed every 30 minutes post infusion for 3 hours"). Use of this verb tense implies that although a particular action was planned, execution of the action may not have been performed to perfection. This verb tense is not susceptible to the inaccuracies resulting from an incomplete deviation list.

Results sections of the clinical study report (Sections 10–13) are, without exception, written using past tense.

Title page, synopsis, table of contents, list of abbreviations and definitions, ethics, investigators and study administrative structure, and introduction

Section 1: Title page

The title page of the clinical study report should include the following information, most of which may be taken directly from the title page of the clinical protocol (on the assumption that the protocol has been written following the ICH E6 Guideline – refer to Chapter 5):

- Title of the study – Use the exact title from the clinical protocol.
- Name of the investigational product – Use the most current name. If this differs from the name on the clinical protocol, a brief explanation should be provided in a footnote.
- Indication studied – Use the indication from the clinical protocol.
- Brief description of study design – Use the brief description from the title page of the clinical protocol.
- Name of the sponsor.
- Protocol identification number.
- Development phase.
- Study initiation date – The first date in the statistical database, which likely represents the first subject's first visit, first laboratory specimen, or first evaluation. Institutional Review Board (IRB) dates, informed consent form dates, and investigator signature dates are not considered initiation dates.
- Study completion date – The last date in the statistical database, which likely represents the last subject's last visit, the date of the last laboratory specimen collected, the date of the last subject visit, or an adverse event date.
- Name and affiliation of principal coordinating investigator or sponsor's medical officer – The name of a medical person responsible for the reviewing study procedures and study conduct.
- Name of company/sponsor signatory – The name of someone whom the health authority can contact with questions.
- Statement of compliance with Good Clinical Practices.
- Date of the report, identification of any previous reports of the same study (this may occur if a report was written based on interim analyses, or if the report required amendment to correct erroneous or incomplete information)

Section 2: Synopsis

The synopsis of a clinical study report is a summary of all study results and should be brief (approximately three pages), unless the complexities of the study require more space. ICH E3 provides a suggested format for a synopsis (Appendix IX). Because of the suggested brevity, and for the sake of clarity, a few small tables often enhance presentation.

A synopsis is always written after all sections of the study report have been completed, in contrast with a protocol, in which the synopsis is written first (Chapter 5). Resist the temptation (or pressure from colleagues) to complete this before completing the results text of the study report (ICH E3 Sections 10–13) because writing the synopsis is easily accomplished by modifying text lifted from the results sections. If the synopsis and the text for Sections 10–13 are written concurrently, discrepancies are likely to occur as one piece of text changes and the other must be updated to match. Health authority reviewers look for discrepancies as an indication of fraud, so any differences that might be construed as an attempt to mislead must be rigorously avoided.

Section 3: Table of contents

Every clinical study report should have a detailed and accurate table of contents, which should include page numbers for sections, as well as for tables and figures, and a list and location of appendices. No regulation or guidance specifies how many heading levels should be included in a table of contents. The ICH E3 clinical study report outline (table of contents) is provided in Appendix X. In addition, this appendix contains a suggested outline, which does not follow the ICH E3 outline exactly but which has been found by the authors to provide a logical flow of study report information.

Section 4: List of abbreviations and definitions

A list of abbreviations and definitions should be provided. In the text of the study report, and separately for the synopsis, spell out the word the first time it is used, and follow this with the abbreviation in parentheses. Each time this technique is done, the word and its abbreviation must be included in the list of abbreviations. A general rule of clear writing is to avoid unusual abbreviations and to limit abbreviations to the most essential (eg, DNA). An example of a list of abbreviations may be found in Appendix VI.

Section 5: Ethics

The purpose of this section of the clinical study report is to record compliance with local and international laws during the conduct of human research. Descriptions of ethical conduct are generally included in most protocols and may be modified for the report.

The description of ethical conduct sections in this book assumes that the trial was conducted in compliance with the laws. If that is not the case, the text will need to reflect deviations from the law and provide an explanation. The sponsor's medical officer or other persons who have participated in the clinical trial should have information pertaining to ethical conduct.

Section 5.1: IEC or IRB
The clinical study report requires a statement that the study and any amendments were reviewed by an IEC or IRB. A list of all IECs or IRBs consulted should be given in ICH E3 Appendix 16.1.3 of the study report and, if required by the regulatory authority, the name of the committee chair should be provided.

Section 5.2: Ethical conduct of the study
A statement is required that the study was conducted in accordance with the ethical principles that have their origins in the Declaration of Helsinki. The Declaration of Helsinki is a statement of ethical principles developed by the World Medical Association to provide guidance to physicians and other participants in medical research involving human subjects. The Declaration has been modified and updated several times since the first writing in 1964 [4]. Whether to cite a specific version of the Declaration of Helsinki is a matter of some debate and requires resolution with the company's Regulatory Affairs department. If a decision to cite specific versions is made, the version(s) should reflect the time during which the study was conducted, not the time at which the clinical study report was written.

Section 5.3: Subject information and consent
How and when informed consent was obtained in relation to subject enrollment (eg, at allocation, prescreening) should be described. A sample of the subject consent form used should be provided in ICH E3 Appendix 16.1.3 of the clinical study report. If the consent form exists in more than one language, a sample of the form in each language could be included. Whether to include all translated versions of the informed consent form is based on company policy, as this is not specified in the guidance.

Section 6: Investigators and study administrative structure
The intent of this section of the clinical study report is to describe and list the various participants who administered, conducted, and evaluated the study. In the interest of keeping the reader focused, and in acknowledgment of the fact that a full description and complete list of all these participants in a large study might take up to 100 pages, information should be summarized for this section, and complete details should be provided in appendices. The report should list names, titles, institutional affiliations, and contact information for:

- Investigators for all clinical study sites (unless the list is greater than approximately five investigators, in which case it is acceptable to provide the name of only the principle investigator and refer to Appendix 16.1.4 of the report).
- Any other person, such as a nurse, physician's assistant, clinical psychologist, clinical pharmacist, or house-staff physician who has played a substantial role (it is unnecessary to list every person who has played an occasional role) in the obser-

vations of primary or other major efficacy variables. As with investigators, if this list starts to become unwieldy, most names may be relegated to ICH E3 Appendix 16.1.4 of the report.
- Contract organizations, contact information, and role.
- Central laboratories, contact information, and the types of specimens analyzed.
- Special committees set up by the sponsor to administer the study or evaluate the outcomes, such as steering committees, executive committees, clinical trial supply management, and safety or monitoring committees. These committees should be described and participants should be named.
- Author of the clinical study report and the biostatistician who analyzed the data.

Section 7: Introduction

An introductory section should be only a few pages in length and include a brief description of the name and indication of the investigational product; the rationale for investigation of the product, which might include a description of the target population, treatment used, duration of treatment, primary endpoints, and rationale for the dose(s) selected; the context of the study within the whole of the clinical program; and any agreements/meetings between the sponsor/company and regulatory authorities that are relevant to the study.

The clinical protocol generally has an introductory section that may be used and updated for the study report. If the introduction in the protocol includes information on results of nonclinical and clinical testing, it will most likely be out-of-date by the time the report is written; this information must be updated or removed. For this reason, it is prudent not to include extraneous information relevant to specific study detail in the protocol, as it quickly becomes outdated and incorrect.

Objectives and methods

Section 8: Study objectives

Study objectives are the heart of a clinical trial, as they drive study population, outcome variables, and analyses, all for the purpose of supporting the product's indication statement. The indication statement is used to market the product because it defines the patient population for whom the drug is intended. Therefore, any statement about objectives must be carefully considered.

For the purposes of the clinical study report, the study objectives from the protocol should be restated here exactly as written in the protocol (with the caveat that spelling errors may be corrected). Any changes to the objectives based on a protocol amendment should be stated following the original wording of the objectives so that the original objectives and the amended objectives are clearly portrayed.

Any change to the study objectives during the course of the study should have been documented in a protocol amendment. If the intent of the objectives changed, but this was not documented in an amendment, then this may be stated separately, but it should not change the wording of the objectives; however, it should always be clear that the change was not included in a protocol amendment.

Section 9: Investigational plan

This section of the clinical study report describes the methods used in the conduct and analysis of the trial, and occupies a good deal of space in the report. A number of revisions from the text of the protocol are necessary to transition it into a study report.

The investigational plan may simply be restated from the protocol, with exceptions for major protocol amendments and protocol deviations and verb tenses (Chapter 5). Organization of the protocol may not be appropriate for a study report due to some points requiring clarification. It is acceptable to reorganize this material or to provide clarification on some point that was obscure in the protocol. Changes to the wording of the protocol that affect an understanding of some fundamental concept of the study, however, should be portrayed clearly as clarifications made for the sake of the study report.

Most protocols, at some point during the study, need an amendment. Amendments may consist of simple clarifications to text that is considered confusing or major revisions that change the objectives of the study, the population enrolled, or the outcomes assessed. Amendments are most likely to affect the investigational plan. The result is that, at the time the report is being written, some aspect of the investigational plan has likely changed from the original protocol, and the regulatory writer is faced with trying to explain the intent of the original protocol and the changes driven by the protocol amendments. Two ways of writing the study methods are:

- Original protocol described – with changes to the original protocol documented in Section 9.8 Changes in the Conduct of the Study or Planned Analyses of the Clinical Study Report. This tactic works best for studies that have changed very little over time.
- Most current protocol described – with changes from the original protocol documented in Section 9.8 Changes in the Conduct of the Study or Planned Analyses of the Clinical Study Report. For studies that have undergone a good deal of change to fundamental concepts such as objectives, study populations, or primary outcome variables, a description of the original protocol would be useless. Therefore, a description of the most current protocol probably would provide the most accurate and clear portrayal of study conduct.

As a caveat to the above recommendations concerning methods descriptions, keeping all references to changes in a protocol until ICH E3 Section 9.8 is an organized way of handling these details; however, it may also unintentionally misconstrue important trial elements (objectives, study population, study design, or primary outcomes) if the reader has to wait until the end of the section to learn what changed. A clinical study report is not a murder mystery. The reader will not be thrilled to find out that, after reading 60 pages of technical material, the control group has been eliminated, the dose of the investigational drug has been reduced, and women are no longer eligible for enrollment. Therefore, a brief description of protocol amendments that affect major conceptual details of the study should be inserted in the methods text. Table 3 presents suggested text for changes in major conceptual study details.

Table 3. *Major changes from the original protocol (most current protocol described)*

Change to major concept	Suggested text
Objectives changed: *From:* The primary objective of the study is to compare changes in blood pressure after administration of panacea acetate to changes noted after administration of placebo. *To:* The primary objective of the study is to assess changes in blood pressure after administration of panacea acetate.	The primary objective of the study is to assess changes in blood pressure after administration of panacea acetate. Objectives as stated above reflect changes from Amendment 1, dated 11 October 2006, to the original protocol, dated 6 August 2005. The original protocol objectives were: The primary objective of the study is to compare changes in blood pressure after administration of panacea acetate to changes noted after administration of placebo. The purpose of the change in the primary objective was to reflect elimination of the control group from the study.
Study population changed: *From:* Men and women with Type 2 diabetes *To:* Men with Type 2 diabetes	Inclusion criteria: - Men with Type 2 diabetes (inclusion of women was removed from the original protocol, dated 4 September 2003, by Amendment 2, dated 12 March 2005)
Study design changed: *From:* randomized, parallel, controlled group comparison *To:* single-group assessment of changes over time	This study design was an assessment of changes in blood pressure from baseline to 2 weeks after initiation of study drug administration. (The study design in the original protocol, dated 6 August 2005, was a randomized, parallel, controlled group comparison. Amendment 1, dated 11 October 2006, removed the control group.)
Primary outcome changed: *From:* changes in ejection fraction 12 months after initiation of dosing *To:* 12-month rate of myocardial infarct	The primary outcome variable assessed the 12-month rate of myocardial infarct. (The primary outcome variable in the original protocol, dated 14 December 2001, was assessment of changes in ejection fraction 12 months post initiation of dosing. This outcome was changed in Amendment 1, dated 3 March 2002, because of poor subject compliance with respect to returning for scheduled ejection fraction assessments.)

Section 9.1: Overall study design and plan: description

This section should describe the investigational product under study, study design, population studied, outcome variables (all discussed in further detail in Chapter 5), and duration of the study (in terms of approximate number of months, years, etc). A table of study evaluations (called a study schedule or flow chart) such as that presented in Table 4, or a study schema (see Chapter 5 for a study schema) is generally available in the clinical protocol.

Table 4. Sample study flow chart

Evaluations	Screening	Week				
		Baseline 1	3	5	7	End of study 8
Panacea		X	X	X	X	
Informed consent	X					
Medical history	X					
Physical exam	X					X
Height	X					
Weight	X					X
Vital signs	X					X
ECG	X					X
Laboratory tests						
Hematology	X					X
Clinical chemistry	X					X
Urinalysis	X					X
PAHG		X	X	X	X	X
IAHG		X	X	X	X	X
Photographs		X	X	X	X	X
Serum pregnancy test	X					
Concomitant medications	X	X	X	X	X	X
Dispense study drug		X	X			
Adverse events		X	X	X	X	X

ECG, electrocardiogram; IAHG, Investigator Assessment Hair Growth; PAGH, Patient Assessment Hair Growth

Section 9.2: Discussion of study design, including the choice of control groups

The intent of this section is to describe the scientific and ethical rationale upon which the study design is based. The gold standard – the prospective, parallel group, double-blind, randomized trial – is easy to justify; however, many other study designs may be appropriate based on a variety of factors including the population studied, the indication for use, features of the investigational product, or location of the study. The rationale should be based on sound scientific knowledge coupled with a regulatory strategy, and therefore the regulatory writer will need to work with other team members to develop this section.

Section 9.3: Selection of study population

Selection of the study population refers to the inclusion and exclusion criteria as described and listed in the protocol. Use protocol text here, with modifications for major changes due to amendments or deviations (Chapter 5). Predetermined (as opposed to unplanned reasons that occur during the trial) reasons for removing subjects from the study should be listed, and again, this is generally described in the protocol.

Section 9.4: Treatments

This section describes the investigational drug or biologic tested in the study, any control agent used, and details of allocation to treatment group, rationale for dose selection, administration, and blinding. The protocol generally supplies these necessary details for both the investigational drug and the control product (if applicable) (Table 5).

Table 5. Description of study drug (The information generally is taken directly from the protocol)

- Name of investigational drug
- Formulation for both active ingredients (drug) and inactive ingredients for both the investigational drug and the control (if applicable), form of the product (tablet, capsule, solution, transdermal patch, etc), source of the product (name and address), and any modifications made to a commercially available test or control product
- Strength: Dose and units (if applicable, a placebo control does not have a dose)
- Route of administration: Oral, intravenous, intradermal, sublingual, subcutaneous, etc
- Batch number: If more than one batch has been used, subjects receiving each batch should be identified in Appendix 16.1.6
- Shelf life and storage: This information may be found in the protocol or the investigator's brochure. Storage conditions are generally described in terms of temperature and light.
- Medications used during the study, other than the investigational drug or control product, are called concomitant medications. Restrictions on, or modifications to, any other treatments or concomitant medications used during the study should be described.
- Medications used for a specified time period just before the study, as applicable.

If the study is 'controlled' (has a treatment group against which observations for the investigational drug will be compared), the investigational plan should describe how subjects have been assigned to these groups and how this was executed. In a randomized study, the randomization codes (and the method used to generate this code), subject identifiers, and treatment assignment should be included in ICH E3 Appendix 16.1.7 of the clinical study report. Subjects should be listed by investigational site for a multicenter study.

Randomization is one method of assigning subjects to a group, but depending on study design, groups may be defined based on time (historical controls). In a historically controlled trial, the method of selection of the historical group should be described. Methods used for blinding (or masking) should be described for studies conducted using this aspect of study design (Chapter 5).

Dose of the investigational drug may be selected in two ways, either for the study as a whole or for each individual subject based on demographic characteristics (age, weight, sex, etc) or medical condition. The rationale for the dose selected for the study should already have been described in the study report's ICH Section 7 Introduction, and is based on scientific experience in animals or humans. Rationale for dose selection and timing of dosing for each subject includes procedures used in administration such as dose escalation, specified titrations, and timing in relation to meals or use of concomitant medications or treatments.

Any plans to measure treatment compliance, the term used to refer to each subject's adherence to protocol-specified procedures, should be described. A subject who does not take the study medication at the required time, who skips doses, or who does not return to the investigational site for planned study visits may be considered to be noncompliant. A number of ways to document the level of treatment compliance exist, such as subject diaries, measurement of study drug in blood or urine, or measurement of the number of tablets/pills/capsules consumed over time. Treatment compliance does not refer to actions performed by investigational site personnel. If site personnel do not follow the protocol, this is considered a protocol deviation (Chapter 5).

Section 9.5: Efficacy and safety variables

A detailed description of the outcome variables (Chapter 5) for the study should be described here as described in the protocol. The description may include the following characteristics:

- Separate lists of efficacy and safety outcome variables: These lists will be essential to writing the results sections of the clinical study report, since the number of efficacy (or safety) variables collected should match the number of efficacy (or safety) variables with results.
- List of all other outcome variables that may not fall into the category of efficacy or safety: Pharmacokinetic sampling or other testing may be specified in the protocol.
- Specification of the primary efficacy variable(s) (if applicable): Not all studies have a primary efficacy variable described in the protocol.

- For each outcome variable: Timing of the procedure or collection of a sample, how this was executed, and special instructions that may affect results.
- Scoring systems, ratings, or scales used, and for a multicenter study, methods by which these were standardized across sites.

Most of the procedures used to measure outcomes in a clinical trial reflect what is referred to as standard clinical practice. Standard practice is not rigidly or exactly defined, but a physician currently practicing medicine in the discipline under study will easily be able to define whether or not a procedure is standard. Use of standard practice procedures implies a certain level of trust in the results. Nonstandard procedures are not always accorded that level of trust and if used in a clinical trial, use must be justified.

If any methods used in the study were not part of standard clinical practice, it may be helpful to describe the rationale for use and why standard practices may have been rejected. Methods used that are not standard should be supported by reference to clinical data, publications, guidelines, or an action by a regulatory authority.

Section 9.6: Data quality assurance
Description of data quality assurance and systems are under the purview of the data management and clinical operations groups, so the regulatory writer will need to collaborate with these groups to provide complete information.

The intent of this section of the clinical study report is to lend credibility to the study results by describing methods of data collection that are accurate, consistent, complete, and reliable. Therefore, training sessions, monitoring, data checking and verification, centralized procedures in the case of a multicenter study, audits, and documentation used to control procedures (instruction manuals) should be described. Most, if not all, of this information has already been included in the protocol. Audit certificates (if applicable) should be included in Appendix 16.1.8 of the clinical study report.

Section 9.7: Statistical methods planned in the protocol and determination of sample size
A description of the planned statistical analyses is generally included in the protocol, but by the time the clinical study report is written, these plans have often undergone a substantial degree of change. The regulatory writer needs to collaborate with the statisticians to revise the description so that it accurately reflects the way the results are presented in the study report's ICH Sections 10–12. Major changes to the planned analysis, whether or not documented in a protocol amendment, should be described briefly in the study report's ICH Section 9.8 Changes in the Conduct of the Study or Planned Analyses of the Clinical Study Report.

Section 9.8: Changes in the conduct of the study or planned analyses
A brief description of changes to the protocol as described in protocol amendment(s) should be provided in this section of the study report. Changes in conduct of the study due to a protocol deviation should also be described in the study report's ICH E3 Section 10.2 Protocol Deviations. The text must clearly delineate between changes due to amendments, and changes due to deviations.

Changes to the planned statistical analysis should be described broadly, since extensive detail may be included in ICH E3 Appendix 16.1.9 of the clinical study report. Statistical changes may be documented in an amendment, but many changes occur outside of an amendment. Yet unlike such changes to study conduct, these changes are generally considered legitimate and perfectly acceptable because statistical testing does not generally affect subject care during the study. It is essential that statisticians be allowed to modify testing based on actual study data.

Results

It is impossible for this book or any other to describe all the ways in which the complex and highly individual results of a clinical study may be portrayed. This chapter offers suggestions but the examples used should not be taken by the reader as the only solution for describing results. A few general rules, however, will apply to writing text for ICH E3 Sections 10–12:

- Results for each of the ICH Sections 10, 11, and 12 (and all subsections as described here) must be supported by statistical data for the study under discussion (not other clinical studies, published material, etc). Statistical data may be in the form of statistical tables, data listings, or data from case report forms.
- Text is helpful but a table may be even more beneficial in helping to explain the information. Tables included in the text of the report (also called 'in-text tables') may be duplicates of the statistical tables or may be reduced to focus on some aspect considered to be of major importance. It is acceptable to have both text and table or text and figure for the same data, a convention not allowed in scientific manuscripts.
- All tables described in the text generally are generated by statisticians and programmers and supplied to the writer as either hardcopy or in some electronic format. It is acceptable for the writer to reduce, collapse, merge, or modify a statistical table if the resulting in-text table is helpful. It is probably not acceptable to recalculate numbers for in-text tables unless this is done in collaboration with help from statistical staff and in compliance with the company's standard operating procedures.
- Text precedes tables, and every table should be introduced and enough description provided to allow the reader to easily understand the numbers. It is not necessary to restate every number that is in the table in the accompanying text.

- Group comparisons may be made using statistical analyses if available, or using clinical judgment (if statistical analyses were not performed). Text describing the results should always clearly state whether or not they are based on statistics. Statements of clinical relevance require collaboration with the sponsor's medical officer.
- Final text of the results reflects team consensus. Adequate and complete characterization of a drug's safety and efficacy profile requires collaboration with medical, statistical, and regulatory experts.

Section 10: Study subjects
Section 10.1: Disposition of subjects
The purpose of this section of the clinical study report is to provide an accounting of the number of subjects who were enrolled in the study, who were randomly assigned to each treatment group, who completed the study, and who may have discontinued prematurely (grouped by the reason for discontinuation). Although this seems a simple exercise, accounting for all subjects in the study sets the stage for the datasets analyzed and therefore is essential for understanding study results. In the example provided in Table 6 and Figure 1, the same number of subject were enrolled in the study, but fewer subjects in the control group completed the study. No statistical testing of the differences between groups was performed, but apparently more subjects in the control group did not complete the study because of an adverse event (three subjects in the investigational drug group and nine subjects in the control group).

Table 6. Example of subject disposition table

Disposition	Investigational drug (Group 1)	Control (Group 2)	Total
Enrolled in study	150	150	300
Completed study	143	134	277
Withdrawn prematurely	7	16	23
Adverse event	3	9	12
Death	1	3	4
Lost to follow up	3	4	7

A brief list of the subjects who withdrew, with an exact description of the reason for withdrawal, may be helpful to understanding why this happened. In almost every clinical study report, a small group of subjects stands out because of either the misfortune of serious illness or a preponderance of side effects from some aspect of study participation (such as the investigational drug). These subject stories tend to unfold as the study report progresses and are generally told in full in the safety evaluation portion of the clinical study report as part of a serious adverse event nar-

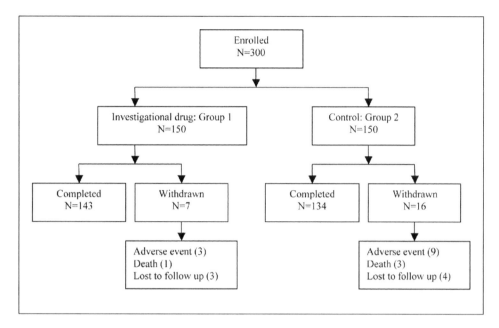

Figure 1. Example of disposition figure

rative. The list of subjects who did not complete the study often contains the beginning of some of these stories.

Section 10.2: Protocol deviations

This section of the study report describes protocol deviations, which are changes to the original protocol in the conduct of the study that have not been described in a protocol amendment (Chapter 5). Deviations range from minor mistakes that have little if any impact on subject safety or the validity of study results to major changes that endanger subjects, invalidate results, or represent intention to commit fraud. Deciding whether a protocol deviation is minor or major is the purview of an entire team. Unfortunately, the decision is not something that can be subjected to mathematical modeling and tends to be the topic of much debate, but a few suggestions may help.

Major deviations may change study outcomes for either safety or efficacy. These deviations generally are departures to enrollment criteria (inclusion or exclusion criteria that affect the study population), incorrect treatment administered or incorrect administration of the investigational drug, or substantial deviations to the timing or execution of study procedures (that affect whether or not the outcome variable has been assessed in such a way as to have any meaning). Failure to obtain informed consent is a major deviation. Speculation on the affect on study results may be stated briefly in this section, but expanded discourse on the topic should reside in the study report's ICH E3 Section 13, Discussion and Overall Conclusions. Major deviations should be listed (either by subject or grouped by type of deviation) and study group. Minor deviations are all deviations not considered to be major and not expected to change study results. Minor deviations are generally not listed specifically.

Section 11: Efficacy evaluation
Section 11.1: Data sets analyzed
The results of every study are analyzed using at least one dataset, or group of subjects, which is defined by statistical analyses. An understanding of the number of subjects in each dataset is fundamental to reporting the results of a clinical trial. Since study outcomes may be analyzed using more than one dataset, they may be reported using one dataset as the primary dataset of interest, with analyses of the remaining datasets used as supporting data only. Deciding which dataset will be used for analysis and reporting of study results is often the result of negotiations with a health authority or team consensus or both. Commonly used datasets are defined in Table 7 (although you may find these definitions modified for a particular study).

Table 7. Examples of datasets

Type	Characteristics
Safety	All subjects enrolled and randomly assigned who received study treatment. Most studies have a safety dataset, which may be used for all outcomes or just for safety outcomes.
Per protocol	All subjects enrolled and randomly assigned in the study who received study treatment and also met a defined set of criteria (number of evaluations performed, minimum time on study, etc). This set is a subset of the safety dataset and is frequently used for efficacy evaluations.
Intent-to-treat	Defined in different ways, but generally all subjects enrolled and randomly assigned, although not all may have received study treatment. Intent-to-treat datasets are generally used for efficacy analyses.

This section of the study report provides an accounting of how many subjects will be discussed in each of the groups analyzed. Table 8 presents an example of three types of datasets (Note: these examples are meant to display a few possibilities but are by no means exhaustive and should not be used to define datasets for any specific study). Note that a total of 303 subjects were enrolled and randomly assigned to treatment groups, and these subjects form the intent-to-treat dataset. Of these 303, three subjects did not receive study treatment and are not included in the safety dataset. A total of eight subjects (three subjects in group 1 and five subjects in group 2) were also excluded from the per-protocol dataset. In addition, statistical analyses may be performed for subgroups within a dataset (for men versus women, for white versus all other races, etc).

Table 8. Example of datasets used

Dataset	Investigational drug (Group 1)	Control (Group 2)	Total
Enrolled	151	152	303
Randomly assigned	151	152	303
Received study treatment	150	150	300
Did not receive study treatment	1	2	3
Intent-to-treat	151	152	303
Safety	150	150	300
Per protocol	147	145	295
Excluded from per protocol	3	5	8
Did not complete study	2	3	5
Did not meet inclusion criteria	1	2	3

Section 11.2: Demographic and other baseline characteristics

The ability to make meaningful comparisons about study groups is at least partly based on the similarity (or dissimilarity) of the groups at the time they enter the study (the baseline period). Demographic characteristics (age, sex, and race or ethnic group) and other baseline characteristics (weight, height, and behavioral or disease or treatment characteristics specific to the indication under study) are used to establish comparability of the study groups.

Table 9 presents an example of a demographic and baseline characteristic table for a clinical trial of the fictional drug estrovex (generic name). The text preceding

Table 9. Example of demographic and baseline characteristics (safety database)

Characteristic	Estrovex 0.3 mg (Group 1) N=150	Control (Group 2) N=150
Age (mean years)	52.9	53.4
Race (N)		
White	134	102
Black	12	25
Asian	2	21
Hispanic	2	2
Age at first menses (mean years)	12.9	12.5
Pregnancies (mean N)	2.6	2.4
Miscarriages, abortions, stillbirths (mean N)	0.5	0.5
Time since last menses (mean years)	2.8	2.7
Moderate to severe daily hot flushes (mean N)	14.5	15.1

Note: for the sake of illustration, this statistical table represents only mean values – standard deviations, standard errors, ranges, and other statistical parameters normally associated with these types of analyses are not shown.

such a table should describe whether the groups were comparable before administration of the investigational drug. Comparability of the groups is the goal of randomization and is desirable because it lends credibility to the results of the study outcomes.

Occasionally, statistical testing of the differences between groups has been performed and this will guide the discussion. If no statistical testing has been done, the comparison is based on clinical judgment, an understanding of the disease being studied, and the affect that differences between groups might have on the study outcomes. In Table 9, no statistical testing has been done. With the exception of race, the treatment groups seem well matched.

Section 11.3: Measurements of treatment compliance

The term 'treatment compliance' refers to each subject's adherence to protocol-specified procedures, not to actions performed by investigational site personnel. If any measures were taken to assess subject compliance, the results of these measures should be described briefly here, by treatment group. If a substantial number of subjects fail to comply with dosing regimens, the validity of study results will be questionable.

Section 11.4: Efficacy results and tabulations of individual subject data

Efficacy outcomes are described in this section of a clinical study report. The number of outcomes described in this section should reflect the number listed in the investigational plan of the clinical study report (ICH E3 Section 9.5). The point of drug (and biologics) research is to describe outcome variables in association with drug administration, either by comparing groups (test group versus a control group) or by comparing time (before administration of the drug versus after administration of the drug). Therefore, in every discussion of results, the relationship to drug dosing should always be a consideration.

The point of the clinical study report is to support the objectives of the study by describing results of the study outcomes, so the first efficacy variable discussed should be the primary endpoint. If a primary endpoint has not been named for the study, then the endpoints considered to be of most value in supporting the objective(s) of the study should be discussed first.

Table 10 presents an example of an efficacy table using the per-protocol dataset. In this example, the study objective is assessment of whether hormone replacement therapy (the fictional drug estrovex) reduces the discomfort of symptoms of menopause. The outcomes (or endpoints) are the number of daily moderate and severe hot flushes at weeks 4, 8, and 12. Care must be taken to use the dataset intended for efficacy evaluations, since there may be more than one. Confer with the statistician responsible for analyzing the study data if in doubt.

In Table 10, statistical testing of the differences between groups has been performed, so results may be stated as being statistically significantly different for the

Table 10. Example of efficacy outcome table (per-protocol dataset)

Moderate and severe hot flushes	Estrovex 0.3 mg (Group 1) N = 150	Control (Group 2) N = 150	P value
Baseline			
Mean number daily	14.5	15.1	0.2556
Week 4			
Mean number daily	8.2	14.6	<0.001
Change from baseline	−6.3	−0.5	<0.001
Week 8			
Mean number daily	7.6	13.2	<0.001
Change from baseline	−6.9	−1.9	<0.001
Week 12			
Mean number daily	5.8	12.5	<0.001
Change from baseline	−8.7	−2.6	<0.001

Note: for the sake of illustration, this statistical table represents only mean values; standard deviations, standard errors, ranges, and other statistical parameters normally associated with these types of analyses are not shown.

groups at weeks 4, 8, and 12. If statistical testing had not been done and if differences between the groups had seemed substantial, the differences would have necessitated a statement about clinically relevant differences (or no clinically relevant differences, depending on the opinion of the medical officer interpreting the data).

Use of the term 'clinically relevant' is somewhat controversial and does not denote any degree of mathematical certainty. Whatever the chosen method is on your team, some way of expressing differences between groups needs to be brought to consensus, as absence of statistical testing is not necessarily equivalent to absence of differences between groups. Waiting for statistical confirmation of every important finding may blind you to trends in the data.

The ICH E3 Guideline lists of number of subsections under Section 11 that describe details of the statistical analysis. These descriptions may best be allocated either to Section 9.7 in the Investigational Plan or to Appendix 16.1.9 of the clinical study report if the description is lengthy. If details of the statistical analyses are necessary to understanding the results, however, brief details of the analytical methods should be provided, with an accompanying interpretation. The statistician on the project may be helpful in interpretation of complex data analyses, and the regulatory writer's role is to work with the medical officer to find language that communicates these statistics in a way that is relevant to clinical medicine.

Section 11.5: Efficacy conclusions

A brief (one to two short paragraphs) and focused summary of the key efficacy results, which includes the primary endpoint (if applicable), should be provided with the goal of tying these results to the study objectives. Results should be stated within the framework of baseline characteristics, protocol deviations, and subject compliance.

Using the estrovex example from Tables 9 and 10, the efficacy conclusion based on demographic data and efficacy outcomes can be written as:

> Daily dosing with estrovex 0.3 mg resulted in statistically significantly greater reductions in the mean number of hot flushes on weeks 4, 8, and 12 compared with the control group ($P<0.001$). Therefore, the primary objective of the study, reduction in frequency and severity of moderate and severe hot flushes after administration of estrovex compared with control, has been met. Racial differences between the groups were noted at baseline (more black and Asian women were enrolled in the control group compared with the extrovex group). No statistical testing of the differences was performed, and whether this difference would affect efficacy outcomes is unknown.

Section 12: Safety evaluation

If study objectives are the heart of a clinical study report, a full and complete description of the safety of an investigational drug is the soul, because it supports subjects' rights to protection from harm by describing and displaying risk. Descriptions of product safety are based on three levels of discussion: (1) extent of exposure (dose, duration, number of subjects); (2) adverse events, laboratory data, vital signs, and physical examinations; and (3) deaths, serious adverse events, and other significant adverse events.

Section 12.1: Extent of exposure

The purpose of describing the extent of exposure is to set the stage for interpretation of safety data. Greater exposure (in terms of duration, dose, or number of subjects) would inherently be expected (although this is not always true) to result in more safety problems. Duration is generally expressed as time (mean or median) but may also be expressed as the number of subjects exposed for specified periods of time and may be broken down by either demographic characteristics (age, sex, race) or disease characteristics. Dose is expressed as a mean or median by dose level (if more than one) and in relation to the number of subjects. Drug concentration is a reference to blood concentrations and may be useful for correlating drug concentration levels with adverse events or changes in laboratory values.

Section 12.2: Adverse events

The analysis of adverse events is generally done using the safety dataset, but on occasion, the analysis may be performed using an intent-to-treat dataset. One important distinction should be made before writing about safety data: does the statistical table display adverse events by number of subjects or by number of events? This difference in results, from number of subjects to number of adverse events, can be dramatic and may alter safety conclusions.

Tables 11 and 12 present hypothetical data for the same subjects, analyzed by number of subjects in Table 11 and number of events in Table 12. Note that in Table 11, the total number of subjects ($N = 418$) is greater than the total number of subjects who experienced an adverse event ($N = 399$), since not all subjects in the study experienced an adverse event. As presented in Table 12, the total number of subjects is the same as in Table 11 ($N = 418$, the total number of subjects), but the number of adverse events is much greater ($N = 544$) because subjects may experience more than one adverse event during the course of a clinical trial. Note that the numbers for death are identical in both tables for the obvious reason that death can only occur once.

In Table 11, a similar number of subjects in both the treatment and control groups appear to have experienced adverse events for all categories except perhaps for death, which appears to have occurred more frequently in the control group. In

Table 11. Example of summary adverse event table by number of subjects

Event	Treatment (N = 208)	Control (N = 210)	Total (N = 418)
	Number (%) of Subjects with Events		
All Subjects With Adverse Events	201 (50.3)	198 (49.6)	399 (100)
Severe	43 (51.1)	41 (48.8)	84 (100)
Treatment-related	54 (49.5)	55 (50.5)	109 (100)
Unexpected	4 (44.4)	5 (55.5)	9 (100)
Serious Adverse Events	45 (50.0)	45 (50.0)	90 (100)
Serious and unexpected	3 (42.9)	4 (57.1)	7 (100)
Treatment-related	20 (50.0)	20 (50.0)	40 (100)
Death	7 (41.2)	10 (58.8)	17 (100)

Table 12. Example of summary adverse event table by numbers of events

Event	Treatment (N = 208)	Control (N = 210)	Total (N = 410)
	Number (%) of Events		
All Adverse Events	248 (45.6)	296 (54.4)	544 (100)
Severe	68 (35.6)	123 (64.4)	191 (100)
Treatment-related	72 (46.2)	84 (53.8)	156 (100)
Unexpected	12 (33.3)	24 (66.7)	36 (100)
Serious Adverse Events	65 (35.1)	120 (64.9)	185 (100)
Serious and unexpected	8 (40.0)	12 (60.0)	20 (100)
Treatment-related serious	22 (35.5)	32 (59.1)	54 (100)
Death	7 (41.2)	10 (58.8)	17 (100)
Withdrawals due to Adverse Events	72 (35.5)	131 (64.5)	203 (100)

Table 12, it becomes apparent that adverse events experienced by subjects in the control group tended to be more severe and were more often serious and unexpected than those in the treatment group. In addition, the number of subjects who withdrew from the study because of an adverse event is higher in the control group (131) compared with subjects in the treatment group (72).

Therefore, the conclusions for these two tables are quite different:

Table 11: Similar number of subjects in each group seemed to have experienced adverse events, both overall (all events) and for the different categories (severity, relationship, and whether expected.)

Table 12: Subjects in the control group experienced adverse events that tended to be more severe, serious, and unexpected.

Statistical testing is seldom performed on adverse event data, so the writer must rely on clinical interpretation of relevance. The overall conclusion is that administration of the investigational drug appears to be associated with a better safety profile than administration of control (or lack of treatment if this is a placebo control).

Section 12.3: Deaths, serious adverse events, and other significant adverse events

Full reporting of drug effects requires that a thorough description be provided for subjects who experience unfortunate events (deaths, and serious and significant adverse events) while participating in a clinical trial. This description should include all deaths and serious and significant adverse events that occurred during the time the subject was in the trial, irrespective of whether or not the event was considered to be related to the study drug under investigation. A thoughtful and comprehensive examination of these events is critical to understanding nuances of drug safety that might be lost in large statistical tables.

An understanding of the difference between serious and severe adverse events is essential. A distinction between the terms is necessitated by differences in the urgency of reporting to the FDA. A serious adverse event is defined as any adverse drug experience occurring at any dose that results in death, is life threatening, requires inpatient hospitalization, is persistent or causes significant disability/incapacity, or causes congenital anomaly or birth defect [5]. Inpatient hospitalization includes initial admission to the hospital on an inpatient basis, even if released the same day, and prolongation of an existing inpatient hospitalization.

All adverse events are reported to the FDA in the clinical study report and certain categories of serious adverse events require reporting within a specified time period after they occur, indicating a degree of urgency (called expedited reporting). A severe adverse event is a designation of a severity rating and has no such reporting requirement. Although many serious adverse events are also severe and so may require expedited reporting, it is possible for a subject to experience a severe adverse event that is not serious (a severe occurrence of allergic rhinitis, for example).

Significant adverse events are more difficult to define. They are not clearly defined by laws or guidelines, and the writer will seldom if ever see significant adverse events on a statistical table. This designation encourages discussion of adverse events that, although not serious, might lead to a better understanding of the safety profile of the drug. Collaboration with the medical officer responsible for the study is essential to determining whether adverse events occurring during a trial warrant discussion as significant adverse events.

A full explanation of deaths and serious and significant adverse events generally requires two types of presentations, tabular summaries and subject narratives. Table 13 provides an example of a serious adverse event tabular summary. The information displayed in an adverse event table varies by drug, since the purpose of the table is to closely examine characteristics that may expose subjects to higher risk. In general, these tables should have a few demographic and baseline characteristics (age, sex, blood glucose, and blood pressure in Table 9), and a brief description of the event. In Table 13, three of four subjects died, all of them women, despite similarities in baseline values.

Table 13. Sample serious adverse event table

Subject Number	Treatment Group	Age (years)	Sex	Baseline Fasting Blood Glucose (mg/dL)	Baseline Blood Pressure (mmHg)	Event
101	LF101	65	F	230	185/130	Acute renal failure, death
103	Control	82	F	180	190/140	Acute renal failure, death
110	LF101	77	F	200	170/120	Acute renal failure, death
115	LF101	82	M	240	200/130	Increased serum creatinine concentration

F, female; M, male

Subject narratives are a description of an individual subject's experience. The format for narratives may either be unstructured (a simple block of text), or may be structured, as in the example in Table 14. Although both structured and unstructured narratives may contain the same information, the structured format allows easier visualization of the characteristics you have selected to show the reader. Although these tables must, by necessity, be modified for each drug, they should generally present subject identifiers (subject ID and initials), site identification (Boston General Hospital), treatment group (LF101), and dose (1 mg), a selected group of demographic and baseline characteristics (age, sex, and baseline blood glucose and blood

Table 14. Sample structured subject narrative

Subject ID/initials: 101/LFW	Site: Boston General Hospital	Treatment group/dose: LF101/1 mg	Date of first dose: 12 November 2007
Age: 65 years	Sex: Female	Baseline blood glucose: 230 mg/dL	Baseline blood pressure: 185/130 mmHg
Adverse event(s):	Acute renal failure, death		
Date of AE: 10 January 2008	Duration: 1 week	Relationship: unknown	Severity: severe

Narrative:
Subject 101, a 65-year-old woman with a history of Type 2 diabetes and hypertension, enrolled in the study on 16 November 2007 and received the first dose of LF101 on 17 November 2007. Baseline serum creatinine value was within normal limits (2 mg/dL), and the subject had no history of renal insufficiency. She continued to receive once-daily doses of LF101 as planned in the clinical protocol for the next 7 weeks. During a scheduled study visit on 10 January 2008, she was noted to have serum creatinine concentrations 3 times the upper limits of normal (6 mg/dL). Despite renal dialysis and supportive care, renal function continued to deteriorate, and she died of acute renal failure on 20 January 2008. This adverse event was considered to be severe and serious, although the relationship to study drug was unknown.

pressure), and basic information about the adverse event (what the event was, the start date, duration, severity, and relationship to the investigational drug).

Section 12.4: Clinical laboratory evaluation
As with results for all outcomes, laboratory results should be compared between groups and over time, with special emphasis on parameters that are known or suspected to be affected by the drug under investigation. Clinical laboratory tests may be displayed a number of ways in statistical tables and tend to be lengthy and complicated to read. Collaboration with the medical officer is essential to finding methods of displaying these data in a meaningful and succinct way, which will often require collapsing tables (cutting rows and columns to display the most important information) before insertion in the text of the report. Use of a collapsed in-text table is perfectly acceptable and should not be confused with eliminating or hiding study data. The full table (uncollapsed) will be included in Section 14 of the clinical study report.

Section 12.5: Vital signs, physical findings, and other observations related to safety
Vital signs, physical findings, and other observations are presented similarly to laboratory variables; that is, they are compared between groups and over time, with special emphasis on findings that are suspected to be affected by the investigational drug.

Section 12.6: Safety conclusions
A brief (one to two short paragraphs), focused summary of the key safety results should be provided to tie these results to the study objectives.

Discussion and conclusions, and appendices

Section 13: Discussion and overall conclusions
This section of the clinical study report should briefly summarize results and conclusions and state whether they support objectives. This section is the only place in the report where it is appropriate to compare the results of the study under discussion with other research (either another study in the same program or published literature). In this way, Section 13 of a clinical study report is similar to the discussion section found in a journal article because it intends to bridge study results to the broader world of clinical medicine.

Any new or unexpected findings should be identified and explored, with implications for clinical use and suggestions for future studies. It is not appropriate in this section to add results that were not discussed and/or displayed in the results sections.

Section 14: Tables, figures, and graphs referred to but not included in the text
All statistical tables and figures are included in Section 14. Clarification of the difference between statistical tables from data listings is based on the fact that statistical tables include summary data (mean, median, standard deviation, etc) of more than one subject. Data listings (also called line listings, raw data) contain individual data points listed for each subject and are included in Section 16 of the clinical study report.

Section 15: Reference list
References cited in the clinical study report are listed here.

Section 16: Appendices
The ICH E3 guidance does not elaborate on contents for the appendices of the study report, and a fair amount of variability seems to exist from sponsor to sponsor without apparent repercussions. The following suggestions are only one solution and should be modified according to logic, good sense, and product needs. Consideration for hard copy versus electronic submissions should be kept in mind, because electronic submissions do not have problems with volumes of paper as hard-copy submissions do.

Several other types of information may be beneficial to include in the appendices. Table 15 provides a list of these materials.

Table 15. Additional materials that may be added to appendices

---
• Grading scales
• Pharmacokinetics report
• Antibody assay report
• Data Monitoring Committee meeting minutes and correspondence

---

Section 16.1: Study information
Section 16.1.1: Protocols
Include the most current protocol, all protocol amendments, and if necessary for the purpose of clarity, all previous versions of the protocol (although including all previous versions would not be the preferred method).

Section 16.1.2: Sample case report forms
Include a sample of the case report form, including only unique pages (to reduce the volume of redundant material, as many pages are identical, or very similar except for the visit number or day).

Section 16.1.3: List of investigators and Institutional Review Boards
List the names and addresses of all investigators and IRB (or IEC, as appropriate). A sample informed consent form should also be included and any other information given to the subject.

Section 16.1.4: Investigators' CV or equivalent summaries of training and experience relevant to the performance of the clinical study
An investigator's curriculum vitae (CV) is considered evidence of the ability to conduct a clinical trial. Due to the volume inherent in many CVs, it is acceptable to make a statement that CVs will be provided upon request for hard copy submissions.

Section 16.1.5: Signatures of principal or coordinating investigator(s) or sponsor's responsible medical officer, depending on the regulatory authority's requirement
A fair amount of confusion exists over signatory responsibility, and differences of opinion exist between regions, complicating this matter further. As a general rule, a clinical study report is presumed to represent statistical data interpreted within the framework of clinical medicine. Therefore, the signature of the medical person responsible for interpreting the data is included on this page. Whether this person is an employee of the sponsor, one of the investigators, or a contract medical officer is not specified in the ICH E3 Guideline and is up to the discretion of the sponsor.

Section 16.1.6: Listing of patients receiving test drug(s)/investigational product(s) from specific batches, where more than one batch was used

A list of batch numbers by subject numbers is helpful in assessing whether or not specific batches may have been associated with drug effects discussed in the study. This information is generally available from manufacturing personnel (Table 16).

Table 16. Sample batch record table

Subject number	Batch number
001	0134
002	0134
003	0135
004	0135
005	0135

Section 16.1.7: Randomization scheme and codes (patient identification and treatment assigned)
A list of subject numbers for each dose group is included. If patient numbers are missing, an explanation is beneficial (Table 17).

Table 17. Sample randomization scheme table

Treatment group	Subject number
0.1 mg CR101	1002 1005 1013 1016
1.0 mg CR101	1001 1004 1007 1008
4.0 mg CR101	1003 1010 1015 1017

Section 16.1.8: Audit certificates (if available) (see Annex IVa and IVb of the guideline)
Any data audit certificates (either internal or external) described in Section 9.6 Data Quality Assurance should be included here. It is not necessary to describe audit results.

Section 16.1.9: Documentation of statistical methods
Because the purpose of the clinical study report is to summarize information and focus on key messages, large quantities of statistical detail are often included in this appendix, with only summarized methods residing in the text of the report.

Section 16.1.10: Documentation of interlaboratory standardization methods and quality assurance procedures, if used

In a study with more than one investigational site, substantiation of comparability of laboratory results may be beneficial. Laboratory data from different sites are generally pooled (ie, the data are pulled together so that statistical analyses, such as means and medians, reflect subjects from all sites) so assurance that laboratory specimens were collected and processed similarly lends credibility to the results.

Section 16.1.11: Publications based on the study

Include all published results of the clinical trial under discussion in the clinical study report.

Section 16.1.12: Important publications referenced in the report

The interpretation of important publications varies greatly. The intent is to supply the reviewer with information necessary to understanding some aspect of the disease under investigation, the methods used for assessments, a statistical method, or any other characteristic of the study that would not be considered general knowledge.

Section 16.2: Patient data listings

Data listings (also called line listings or raw data) are statistical output listing each individual subject's data (Table 18) and are supplied by the statistical group. Each page of data listings presents data of a certain category (such as demographic characteristics) for several subjects and as such data listings are considered to be organized by 'variable'. Data listings organized by 'subject' are described in ICH E3 Section 16.4.

Table 18. Sample data listings for demographic characteristics

Listing 16.1.1 Demographic characteristics by subject							
Subject no.	Initials	Treatment	Race	Age (years)	Sex	Blood pressure (mmHg)	Weight (kg)
1001	MD	LF101	White	65	F	180/120	68
1002	FM	Placebo	White	77	F	200/124	73
1003	GHR	LF101	White	82	M	154/100	80
1004	LES	LF101	Asian	56	M	178/130	78
1005	WWA	Placebo	White	76	M	180/134	81
1006	RK	Placebo	White	45	F	190/120	70

F, female; M, male

Section 16.3: Case report forms

Case report forms are the forms (paper or electronic) on which all human data are recorded by the investigational site personnel. The purpose of submitting case report forms is so that health authority personnel may see for themselves what was written on the forms. Inclusion of case report forms in a study report requires consideration of the type of submission: hard copy or electronic. Each subject has a set of forms that may be up to 300 pages for a complex study. Few math skills are required to estimate the massive volume of paper this could generate for a submission from a large clinical trial. Your regulatory affairs department may have negotiated the method of case report form transfer to avoid the potential volume of paper. Often, only case report forms for subjects who have died or experienced serious adverse events are included. An electronic submission has no such problems.

Section 16.4: Individual patient data listings

Individual patient (or subject) listings differ from the data listings required in ICH E3 Section 16.2. Individual listings are organized by subject, and each page presents data for several variables for one subject (Table 19).

Table 19. Example individual subject data listing

Listing 16.1.1 Subject 1001							
Subject no.	**Initials**	**Treatment**	**Race**	**Age (years)**	**Sex**	**Blood Pressure (mmHg)**	**Weight (kg)**
Baseline							
1001	M-D	L101	White	65	F	180/120	68
	Con Meds.	**Informed Consent**	**Preg. Test**	**Pain Score**			
	Acetaminophen Prednisone	Yes	Neg	86			
Visit 1	Dosing LF101			**Pain Score**		**Blood Pressure (mmHg)**	
	Yes			90		182/124	
Visit 2	Dosing LF101			**Pain Score**		**Blood Pressure (mmHg)**	
	Yes			72		165/90	

Abbreviated reports

An abbreviated report should contain a full report of safety information, as described in the ICH E3 Guideline. It should also contain enough efficacy information to allow the reviewers to fully assess whether the efficacy results, if any, cast doubt on the effectiveness of the investigational drug for the proposed indication. It is often sufficient to write a small section on the primary efficacy variable, include an in-text table, and refer the reviewer to the statistical tables and listings for further information.

As presented in Table 20, the outline of an abbreviated study report should contain only selected sections of ICH E3. These sections were described earlier in this chapter. As noted for a full ICH E3 report, there is no need to follow this outline in a rigid fashion. It is meant to be a suggestion of content.

Table 20. Sections to be included in an abbreviated study report (numbered as described in ICH E3)

- Section 1 – Title page
- Section 2 – Synopsis
- Section 3 – Table of contents for the individual clinical study report
- Section 4 – List of abbreviations and definitions of terms
- Section 9.1 – Overall study and design and plan: description
- Section 9.8 – Changes in the conduct of the study or planned analyses
- Section 10.1 – Disposition of patients
- Section 12 – Safety evaluation
- Section 13 – Discussion and overall conclusions
- Section 14 – Tables, figures and graphs referred to but not included in the text
- Section 16.1.1 – Protocol and protocol amendments
- Section 16.1.2 – Sample case report forms (unique pages only)
- Section 16.3.1 – Case report forms for deaths, other serious adverse events and withdrawals for adverse events (submit under item 12 – FDA form 356h)
- Section 16.4 – Individual patient data listings for safety data. Individual patient listings of efficacy data are not necessary.
- A summary of the efficacy evaluation (suggested to be primarily in table form).
- The summary should contain enough information for the reviewer to determine whether the study results are germane to the overall evaluation of effectiveness and to use in review of the integrated analysis of effectiveness, if necessary (including means, confidence, intervals, p-values, standard errors, etc). Section 11.4.1 of ICH E3 format may be used, if appropriate.
- Any additional information pertinent to the evaluation of safety should also be included.
- Section 12, Safety evaluation, should provide comprehensive safety information. Other sections should be concise and need not be as comprehensive as in a full report.

Synopsis

A synopsis should contain sufficient information to allow the reviewer to assess whether the results of the study cast doubt on the safety of the investigational drug for the proposed indication. Appendix IX includes a sample clinical study report synopsis structured as suggested by ICH E3. The FDA considers this is an acceptable format to use. A study protocol and protocol amendments should be appended with the synopsis.

The guidance for abbreviated reports allows a good deal of flexibility in terms of the level of detail required for the safety discussion in a synopsis (as brief as in a full study report, or fully expanded as in Section 12 of a full report). In general, a complete discussion (as in Section 12 of the ICH E3 Guideline) is written, but this may be included in the synopsis or in the Integrated Summary of Safety (Chapter 9). Published literature, with appended safety data, may be submitted instead of a synopsis.

Side bar: Lessons learned

In all controlled studies in which the investigational product is being compared with either a placebo control or another product, the writer is always challenged to find ways to describe group differences. Statistical testing makes this relatively easy, as the writer may rely on stating that X either is or is not statistically significantly different from Y. Negative statistical results (no differences), or an absence of statistical testing, makes the description of comparisons more difficult.

The word 'significance' tends to create controversy because in the presence of a modifier such as 'statistical' it connotes a high (but not always deserved) level of confidence in results. But when modified by the word 'clinical,' it often loses this level of assurance, as lack of mathematical certainly is associated with whim (and also not always deserved). So in the absence of statistical testing, and to avoid the word 'clinical' which may not be seen as a 'hard' measure, the word 'significance' is sometimes used without any modifier. This is undeniably the worst solution, as now all information about statistical testing has been eliminated (Was it done? Was the result statistically significant?), and no certainty about clinical relevance has been provided. A better solution is to use the word 'clinically' if, in the opinion of the medical officer, the results are clinically relevant (which is perfectly appropriate, as clinical relevance is not always measurable) or, if in doubt, the word 'substantial' may be used. In drug and biologics research, this word has no defined meaning, so it may be used to describe something that differed in a way that mattered, without endless discussions of clinical relevance. A word of caution: the word 'substantial' has a defined, legal meaning in medical devices and should not be used unless sanctioned by your regulatory team. It is never used to describe clinical outcomes.

References

1 ICH E3, Guideline for Industry, Structure and Content of Clinical Study Reports, July 1996.
2 Guidance for Industry, Submission of Abbreviated Reports and Synopses in Support of Marketing Applications, U.S. Department of Health and Human Services, Food and Drug Administration, Center for Drug Evaluation and Research (CDER), Center for Biologics Evaluation and Research (CBER) August 1999. http://www.fda.gov/cber/gdlns/abbrev.htm. (Accessed 15 April 2008).
3 International Committee of Medical Journal Editors. Uniform Requirements for Manuscripts Submitted to Biomedical Journals: Writing and Editing for Biomedical Publication. Updated October 2005. Available at http://www.icmje.org. (Accessed 11 January 2006).
4 The World Medical Association, Policy, World Medical Association Declaration of Helsinki: Ethical Principles for Medical Research Involving Human Subjects. http://www.wma.net/e/policy/b3.htm (Accessed 27 March 2007).
5 Title 21 – Food and Drugs Chapter I – Food and Drug Administration Department of Health and Human Services. Subpart B – Investigational New Drug Application (IND) §312.30 IND safety reports. http://www.accessdata.fda.gov/scripts/cdrh/cfdocs/cfcfr/CFRSearch.cfm. (Accessed 30 March 2008)

Integrated documents

Targeted Regulatory Writing Techniques. Clinical Documents for Drugs and Biologics,
edited by Linda Fossati Wood and MaryAnn Foote
© 2009 Birkhäuser Verlag Basel/Switzerland

Chapter 7.

Investigator's brochures

Linda Fossati Wood

MedWrite, Inc., Westford, Massachusetts, USA

Introduction

All marketed drugs and biologics are sold accompanied by the package leaflet (European Union [EU]) [1] or a package insert (Japan and the United States) [2–5], documents that describe product characteristics (active and inactive ingredients, chemical structure, formula, and pharmaceutical properties), summarize all known nonclinical and clinical information, and provide guidance for dosing and administration. The contents of these 'labels' represent the culmination of all research and development testing and of negotiations with the health authority from whom marketing approval is requested. For a marketed product, labeling documents constitute the primary method of communication with the physician prescribing the product.

Package leaflets and package inserts do not exist in the pre-approval stage of drug and biologics development because the research necessary to write an insert is in progress. Therefore, the function of communicating all known product information is under the purview of a document called an investigator's brochure, which contains a compilation of all known nonclinical and clinical information essential to use of the product in humans. During the course of product development, as additional studies are performed and the product's safety and efficacy profile are better characterized, the investigator's brochure goes through substantial changes and eventually provides the basis for labeling.

As with all written communication, an understanding of the audience is essential. The audience for an investigator's brochure is clinical investigational site personnel, people actively involved in clinical medicine, who are generally found in hospitals or clinics. They are seldom in a position to sit and read, but they are directly in contact with study subjects so they must be thoroughly familiar with a product's characteristics to ensure safe use. Therefore, a good investigator's brochure is brief (approximately 50 pages for the United States; see side bar for more information concerning investigator's brochures in Japan), focuses clearly on details most critical to subject safety and potential benefits, and refers the reader to additional materials available from the sponsor if expanded information is desired. Use of bulleted lists, tabular displays, and figures is particularly helpful.

Investigator's brochures are written using the ICH E6 Guidance [6] and in close collaboration with a multidisciplinary team that usually includes medical, nonclinical, manufacturing, and regulatory expertise and possibly other groups at a company. The guidance contains an outline of suggested contents in three fundamental areas:

- Nonclinical: testing in animals
- Clinical: testing in humans
- Drug description and chemical or biologic characteristics: formulation, dosing, administration, and storage information

It is not necessary to follow this outline exactly. Instead, the organization should always be dictated primarily by logic and good communication principles, but careful consideration should be given to including all content if possible. This chapter describes writing an investigator's brochure in terms of these areas of content and describes changes in the balance of nonclinical and clinical information during the life cycle of the brochure.

Side bar: Lessons learned

When preparing investigator's brochures for use in Japan, it is important to know that this document is used somewhat differently in Japan compared with Europe and the United States. In Europe and the United States, the investigator's brochure is considered to be the beginning of the package insert for a marketed product. The investigator's brochure is quite an extensive package insert. It includes a large amount of nonclinical data because clinical data on the product or the indication are limited in the early stages of development. As more human data are collected, the usual procedure is to reduce the volume of nonclinical data, either by additionally summarizing or completely eliminating some of the nonclinical studies. This procedure is modified in Japan, where the regulatory agency uses the investigator's brochure almost like an IND. Nonclinical data are not removed, and as new studies are added, the brochure can become larger. Many companies that develop drugs globally place the nonclinical data in appendices to the brochure. In this way, the main body of the brochure is the same globally – important because it is labeling and a product must be labeled consistently – but the need for all nonclinical data, particularly animal toxicology studies, is met for the Japanese authorities.

Title page, table of contents, list of abbreviations, summary, introduction

Title page
The title page includes important information, primarily related to administrative functions (Table 1).

Table 1. Title page of an investigator's brochure

- Sponsor's name
- Name of the investigational product – the most current name and any additional names that may be known by the investigators that help identify the product
- Research number – an additional product identifier that may or may not exist at every company
- Edition number – identifies versions of the investigator's brochure. The first version finalized and sent to the investigational site would be version 1. Version identification is done for the purpose of determining what information was in the possession of clinical site personnel at a specified time point, which is not the same as a version number used by word processing software and should not be confused with the multiple versions circulated internally within the team for the purpose of review
- Release date – date the document is finalized and signed off; the effective date of the information
- Replaces previous edition number – sequentially number versions allow easy version control and tracking
- Date – date of the previous version; this information is not needed for version 1
- Safety data cut-off date – date of the last safety report

Table of contents, table of tables or figures
Every investigator's brochure should have a detailed and accurate table of contents, which should include page numbers for sections as well as for tables and figures, and a list and location of appendices. No regulation or guidance specifies how many heading levels should be included in a table of contents. An example of a table of contents for an investigator's brochure is provided in Appendix XI.

List of abbreviations
A list of abbreviations and definitions should be provided for the purpose of defining abbreviations used in the document. In the text of the investigator's brochure, spell out the word the first time it is used, and follow this with the abbreviation in parentheses. Each time this technique is done, the word and its abbreviation must be included in the list of abbreviations. An example of a list of abbreviations is given in Appendix VI. Caution should be exercised in using nonstandard abbreviations and too many abbreviations.

Summary
The summary of an investigator's brochure is a summary of all three content areas contained in the document: drug description and chemical or biologic characteristics, nonclinical testing, and clinical testing (if applicable; the first edition of an investigator's brochure often does not contain any results of human testing). The

summary should not exceed two pages, unless the complexities of the product require more space.

A summary should always be written after all sections of the investigator's brochure have been completed, although in the rush to meet deadlines, this rule is often a ready casualty. Writing the summary is easily accomplished by modifying text from other sections, and fewer discrepancies are likely to result as one piece of text changes and the other must be updated to match.

Introduction
A brief introductory statement includes the following information:

- Product description: chemical and generic name, all active ingredients, the pharmacological class
- Any advantages the investigational product is expected to have in this class
- Indication: the disease, syndrome, or diagnosis for which the product is intended and the population for whom it is intended (Chapter 5, Protocols). The indication statement, generally the result of cross-functional team consensus, requires that the writer work with regulatory affairs to determine the status and exact wording.
- Investigational plan: a very brief (one to two sentences) description of clinical development plans. Clinical development plans comprise the number and types of clinical studies planned for the coming year. Care should be taken not to elaborate excessively on this point, since it will likely change substantially over the course of development.

Physical, chemical, and pharmaceutical properties and formulation

This section is a very brief (one to two pages) description of drug characteristics, sufficient to orient the investigator to fundamental aspects that help to place the investigational product in a therapeutic class and assist in predicting potential drug effects. The information used in this section is provided by manufacturing staff (Table 2).

Table 2. Information to be contained in physical, chemical, and pharmaceutical properties and formulation section

---

- Drug substance (the active ingredient, which is used to make the drug product) description: chemical and or structural formula, relevant physical, chemical, and pharmaceutical properties
- Drug product (the active ingredient plus all inactive ingredients) description: the formula, including all inactive ingredients, with instructions for storage and handling. Drug product is administered to subjects.
- Structural similarity to other known compounds (if applicable)
- Dose, route of administration
- Handling and preparation for administration (if applicable)
- Storage

During the course of development, certain manufacturing and testing character-istics or procedures are expected to change as knowledge of the specific formulation improves. Although these changes are anticipated as a fundamental aspect of drug development, any change may potentially alter drug effects on biologic systems. Be-cause testing in animals intends to predict drug effects in humans, use of the same batch of drug product for both nonclinical and clinical testing (which is assumed to have been manufactured and tested under identical conditions using identical ma-terials) is optimal. Because it may not be possible to do such testing, any substantial difference between the drug product used in nonclinical testing and that to be used in clinical testing should be briefly described and presented in a table. Table 3 pres-ents an example of a change in formulation table, but in reality the content of the tabular presentation will vary widely from product to product.

Table 3. Sample nonclinical and clinical final formulation differences

Drug product	Nonclinical studies		Clinical study (Phase 1)
	Single-dose toxicity (101, 201)	Multidose toxicity (001, 002)	
Lot no.	E4567	E3478	E3478
Drug (mg/mL)	30	15	15
pH	5.6	4.6	4.6
Impurities	8%	3.2%	3.2%

Nonclinical studies

At an early stage in development, nonclinical information generally forms most of the argument for safe first use in humans. After initiation of clinical trials, and as human data become available for inclusion in the investigator's brochure, the rele-vancy of certain nonclinical studies may diminish somewhat, and consequently these sections should be culled to retain the size of the brochure (approximately 50 pages) as clinical trial data are added and those sections expanded. Nonclinical research is generally divided into three categories, each of which generally has associated study reports or publications:

- Pharmacology
- Pharmacokinetics and metabolism
- Toxicology

As presented in Figure 1, this section comprises both integrated text (discussion of more than one study) and text from a source document (eg, an abstract discussing results of a single study). Summarized nonclinical information may be structured

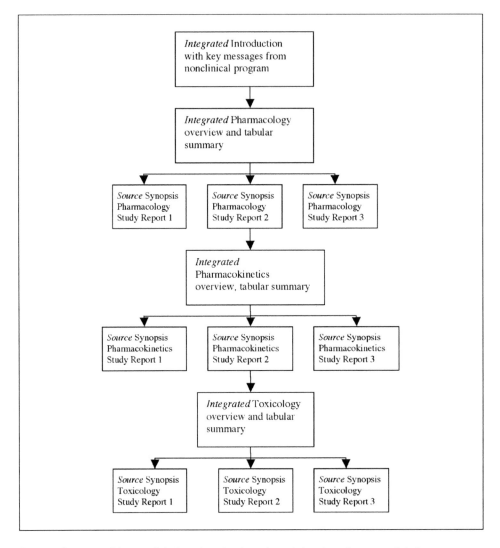

Figure 1. Structure of the nonclinical section of an investigator's brochure: large nonclinical program

using an introduction with key messages, overview text, tabular summaries, and synopses from the individual study reports. Abstracts from publications may be used in place of a study report synopsis. Note that information in the introduction is the most summarized in the nonclinical section, and that the level of detail expands in complexity and volume in the source text. This structure is appropriate for nonclinical programs that are extensive, and include a large number of studies.

Very often, the nonclinical section of the investigator's brochure is written by scientists in the appropriate departments and provided to the medical writer. In this situation, the medical writer should edit the material to fit the company style so the final document appears to have been written in 'one voice' by one writer.

Organization of nonclinical information in the investigator's brochure as presented in Figure 1 distinguishes between the three nonclinical categories (pharmacology, pharmacokinetics, and toxicology) and provides a tabular summary and overview text individually for each of these three categories in an effort to organize a large amount of information. In a smaller program with fewer studies, it may be beneficial to simplify organization and summarize all studies in one tabular summary, and in one overview summary (Figure 2).

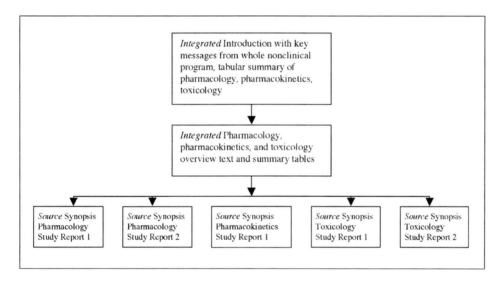

Figure 2. Structure of the nonclinical section of an investigator's brochure: small nonclinical program

Introduction in nonclinical section

The point of this section is to introduce the reviewer and clinical trial staff to the investigational product and its use for the proposed indication. A few brief statements about the nonclinical program (species tested, routes of administration, acute versus chronic testing, doses), and a short bulleted list of key messages that sharply focuses on a few of the most relevant points help to orient the reader to the program. Relevancy is based on communicating characteristics of the drug or biologic that will allow a clinician to most safely and effectively use the investigational product.

All three study categories (pharmacology, pharmacokinetics, and toxicology) should be mentioned in the key messages, as shown in Table 4 for the fictional drug xerimax, under development for an oncological indication. The first point (pharmacology) makes a statement about the effects of the drug xerimax on tumor growth. This statement is therefore intended to support efficacy (or in the case of oncology drugs, activity). The second key message (pharmacokinetics) states observed metabolic characteristics. The third point (toxicology) describes safety considerations that may help to predict adverse events in humans (also known as side effects).

Table 4. Example of key messages

• Xerimax 100 mg/kg IV showed a 223% delay in tumor growth compared with untreated controls in the A2780 human ovarian cancer xenograft model in mice
• C_{max} for a 17 mg/m^2 dose of xerimax was 655 ng/mL, and $t_{1/2}$ was 4.15 hours
• Xerimax was well tolerated in rats in doses up to 1200 mg/kg IV and in dogs in doses up to 30 mg/kg, with only minimal hematopoietic suppression in both species (neutropenia and thrombocytopenia) and reduced food consumption in dogs
---

Overview in nonclinical section

An overview is text that integrates information from several studies. For the purposes of an investigator's brochure, the overview sections for pharmacology, pharmacokinetics, and toxicology should each be approximately one to five pages (with the exception of a small nonclinical program, in which the integrated overview of pharmacology, pharmacokinetics, and toxicology is combined for a total of approximately one to ten pages), and provide a synthesis of the most relevant information (the term 'relevant' is not defined, but allows for elimination of those studies that do not substantially contribute to an understanding of the drug due to selection of a poor model, dosing or procedural deficiencies, lack of data collection, etc).

Tabular presentation may be beneficial to understanding drug characteristics for the most important assessments in light of differences in dosing, species used, study design, or outcomes assessed. Again, tabular presentations should be modified to suite the product. In Table 5, the tabular presentation notes the differences between two studies with respect to the study design (dose and frequency of dosing, time to endpoint and neurological assessments), and results (% tumor growth delay, time to endpoint, ie, growth of the tumors to a specified size, and toxicity).

Table 5. Sample nonclinical overview tabular presentation of relevant outcomes across more than one study

Outcomes/results	Study 1 Xerimax 100 mg/kg IV/ week × 3	Study 2 Xerimax 50 mg/kg IV/ week × 6
Species		
Athymic nude mice (N)	50	98
Model	Human xenograft LS174T colon CA	Human xenograft A2780 ovarian CA
Major outcomes		
%Tumor growth delay	At time to endpoint	At time to endpoint
Time to endpoint (days)	Tumors 1500 mm^3, mice euthanized	Tumors 1200 mm^3, mice euthanized
Toxicity	Hematology	Hematology
	Neurotoxicity	Not collected
Results		
% Tumor growth delay		
Xerimax	223	250
Positive control (trexon)	120	90
Time to endpoint (mean days)		
Xerimax	59	60
Positive control (trexon)	45	50
Toxicity		
Hematopoietic	Well tolerated	Well tolerated
Neurotoxicity	Increased Rotarod day 5	Not applicable

Tabular summary

A tabular summary is a table containing a summary of all relevant nonclinical pharmacology, toxicology, and pharmacokinetics, and investigational metabolism studies. This presentation should provide an easy, quick, and focused look at the entire nonclinical program to date. Contents for a tabular summary will by necessity vary somewhat from one program to the next, but species, the number of animals studied (N), the dose and route of administration, the number of doses, the model used, major endpoints, and results of major endpoints should be displayed. Each study occupies one row of the table (Table 6).

Table 6. Sample nonclinical studies tabular summary

Species	N	Dose (mg/kg)	Methods	Results
Pharmacology studies				
Study 1				
Athymic nude mice	10 F/ group Total 50	**Control:** • Group 1: untreated • Group 2: trexon 100 IV **Test:** • Group 3: xerimax 100 IV	• Dosing once weekly × 3 • LS174T human colon carcinoma xenograft tumors established • Time to endpoint = tumors 1500 mm^3, mice euthanized • Activity = % tumor growth delay	**Control group tumor growth delay:** • Group 2: 120% **Xerimax tumor growth delay:** • Group 3 = 223% • Well-tolerated
Study 2				
Athymic nude mice	14 F/ group Total 98	**Control:** • Group 1: untreated • Group 2: trexon 100 IV **Test:** • Group 3: xerimax 50 IV	• Dosing twice weekly × 6 • A2780 human ovarian cancer xenograft tumors established • Time to endpoint = tumors 1200 mm^3, mice euthanized • Activity = % tumor growth delay	**Control group tumor growth delay:** • Group 2: 90% **Xerimax tumor growth delay:** • Group 3 = 250% • Well-tolerated

Synopses in nonclinical section

A well-written synopsis from an individual study report (or the abstract from a publication) should be sufficiently brief and informative to use for the investigator's brochure. If an investigator requires the full text of a study report, he or she may request it from the sponsor, so there is no need to provide extensive details in the investigator's brochure. Occasionally, a nonclinical study spans the different study categories (pharmacology, pharmacokinetics, toxicology). If synopses are sufficiently summarized, it may be appropriate to put the same synopsis in twice, for ease of finding information. A preferred solution would be to reference another section in which the synopsis may be found.

The nonclinical synopses included in this section of the investigator's brochure should be those for the most relevant studies in the tabular summary. Several studies listed in the tabular summary may be of lesser importance to an understanding of the investigational product and so may not be considered of sufficient importance to increase the bulk of the brochure.

Effects in humans

The first version of any investigator's brochure is unlikely to have information on the effects of the investigational product in humans when it is submitted with a Clinical Trial Application (CTA, Europe), Clinical Trial Notification (CTN, Japan), or Investigational New Drug Application (IND, United States), which represent the first request for use in humans. Clinical testing or marketing may have been initiated in another country or may occurred in the past as part of a previous development program, in which case this section should provide a summary of all known information. As time progresses and clinical trial data become available, this section expands and nonclinical sections tend to shrink, except for investigator's brochures used in Japan.

This section comprises both integrated text (discussion of more than one study) and text from a source document (discussion of a single study) (Figure 3). The structure for clinical information may be similar to that for nonclinical information for a small nonclinical program, using an introduction with key messages, overview text, tabular summaries, and synopses from the individual study reports (or abstracts from publications). Clinical data are generally categorized by

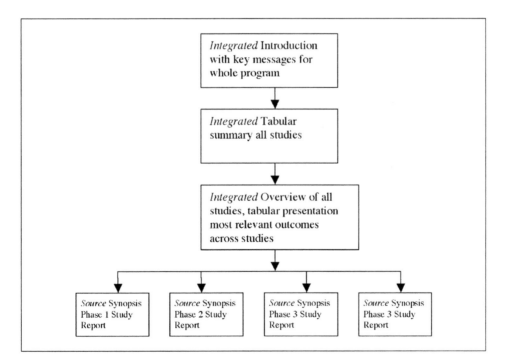

Figure 3. Structure of the clinical section of an investigator's brochure

study phase (Chapter 1, Developing a Target), starting with phase 1, and continuing on to phases 2, 3, and 4 rather than physiologic effect as in nonclinical studies (pharmacology, pharmacokinetics, and toxicology). In general, many fewer human studies tend to be conducted compared with nonclinical studies, so structure of this section tends to be simpler and more similar to that for a small nonclinical program.

Introduction to use in humans

A few brief statements about the populations tested in all clinical studies to date (volunteers versus subjects with the disease, men and women, disease characteristics or indications, age groups, etc), region in which the testing was conducted, routes of administration (by mouth, intravenously, subcutaneously, etc), doses (using quantity and units), and a short bulleted list of key messages that focus on product characteristics that are relevant to safe and effective clinical use should be in the introduction. Key messages should support the planned indication for use, and are developed in collaboration with the medical, regulatory, and statistical staff.

Key messages in clinical trials are generally categorized as:

- Efficacy (or activity in oncology trials)
- Pharmacokinetics/pharmacodynamics
- Safety

In the example for the fictional drug xerimax (under development for an oncological indication), bulleted points shown in Table 7 focus on efficacy (the first two key messages), pharmacokinetics (the third key message), and safety (the last two key messages) of both the investigational product (xerimax) and an active control agent (trexon, also fictional).

Table 7. Example of key messages

- Xerimax 10 mg/kg IV every 3 weeks for 6 cycles was associated with a 25% improvement in overall survival (24 months) compared with the active control, trexon (19 months, $P=0.0152$) in phase 2 study 003.
- Progression-free survival was improved for subjects who received xerimax 10 mg/kg versus subjects who received trexon (9 months for subjects who received xerimax versus 6 months for subjects who received trexon, $P=0.0110$).
- In a phase 1 study of xerimax 15 mg/kg IV every 3 weeks for 3 cycles, C_{max} was 106 ng/mL, and $t_{1/2}$ was 6 hours after dosing in cycle 1.
- Xerimax was associated with grade 1 and 2 hematopoietic suppression (neutropenia and thrombocytopenia) and occasional grade 1 diarrhea occurring on day 15 of cycle 1 in both phase 2 studies. Signs of toxicity did not increase with additional cycles, and all subjects enrolled completed 6 cycles.
- Active control (trexon) in a phase 3 study was associated with grade 2 and 3 hematopoietic suppression (neutropenia and thrombocytopenia), and grade 4 diarrhea in 3 subjects. Eight of 22 subjects in this treatment group (36.4%) withdrew before completion of 6 cycles because of toxicity.

Overview of human experience

An overview is text that integrates information from several studies. For the purposes of an investigator's brochure, the overview clinical section should be fewer than 10 pages and provide a synthesis of the most clinically relevant information. Determination of clinical relevancy is done in collaboration with a medical officer.

Tabular presentation of important outcomes across all clinical studies (or those considered to be most representative of the product's characteristics) may be beneficial. Again, tabular presentations should be modified to suite the product. Results of three individual phase 1 and phase 2 studies are presented in Table 8, allowing comparison of the efficacy and safety observed in all three studies and for results of pharmacokinetics from the phase 1 study. In later stages of development, at the time clinical trials have been completed and the sponsor is preparing to submit for marketing approval (an MAA in Europe and Japan, and an NDA in the United States), data from all clinical trials are generally merged into one database, and tables such as these are generated by the statisticians. In early development stages, when few human trials (if any) have been conducted, the regulatory writer may have to compile the tables manually.

Table 8. Sample clinical overview tabular presentation of relevant outcomes across more than one study

	Phase 1	Phase 2	
	Study 001￼N=12	Study 002￼N=20	Study 003￼N=24
Xerimax dose and regimen	5–20 mg/kg IV q3 weeks × 3	10 mg/kg IV q3 weeks × 6	10 mg/kg IV q3 weeks × 6
Indication	Colorectal CA	Colorectal CA	Ovarian CA
Efficacy	Mean	Mean	Mean
Overall survival	NA	22.5 months	24 months
Progression-free survival	NA	7 months	9 months
Pharmacokinetics (cycle 1)			
C_{max}	106 ng/mL	NA	NA
$T_{1/2}$	6 hours	NA	NA
Safety (last cycle)			
Hematology			
Hemoglobin concentration	9.6 g/dL	11.2 g/dL	12.6 g/dL
Hematocrit	30%	34%	36%
Absolute neutrophil count	1100 cells/mm^3	1245 cells/mm^3	1055 cells/mm^3
Platelet count	70,000 cells/mm^3	86,000 cells/mm^3	92,000 cells/mm^3
Withdrawn due to adverse event	0 (0%)	3 (15%)	2 (8.3%)

CA, cancer; IV, intravenous; NA, not applicable; q3, every 3 weeks

Tabular summary

The tabular display in Table 8 presents a focused look at the most important efficacy results from more than one study. This type of display is meant to allow easy comparison and discussion of outcomes (overall survival, for example) and should not be confused with a tabular summary, which has a slightly different purpose (Table 9).

Table 9. Sample clinical studies tabular summary

Study no. (Country)	N	Methods	Results
Phase 1			
001 (UK)	12	Phase 1 open-label, first time in subjects with colorectal cancer, IV dose-escalation study (xerimax 5, 10, 15, 20, 25 mg/kg) every 3 weeks × 3, until maximum tolerated dose	Maximum tolerated dose = 10 mg based on grade 3 neutropenia and thrombocytopenia
Phase 2			
002 (US)	20	Phase 2 open-label study of xerimax 10 mg/kg IV every 3 weeks × 6 in subjects with colorectal cancer, evaluated overall survival and progression-free survival, safety.	Progression-free survival = 9 months Overall survival = 24 months Well tolerated, grade 2 diarrhea in several subjects, no hematopoietic suppression, no deaths
003 (US)	24	Phase 2 open-label study of xerimax 10 mg/kg IV every 3 weeks × 6 in subjects with ovarian cancer, evaluated overall survival and progression-free survival, safety.	Progression-free survival = 6 months Overall survival = 14 months Well tolerated, grade 3 neutropenia several subjects, no serious adverse events, no deaths

A tabular summary is a table containing one row each for all clinical trials (unlike nonclinical studies for which only relevant studies may be listed, all clinical trials in humans should be listed here). The purpose of a tabular summary is to provide an accounting of all clinical trials conducted (the number of trials, the number of subjects, the regions in which the trials were conducted, the progression from one phase to the next, etc), as well as providing an easy, quick, and very focused look at the entire clinical program to date. Contents for a tabular summary will by necessity vary, but the number of subjects studied (N), the dose and route of administration, the number of doses, major disease characteristics, timing and results of major endpoints should be displayed. Each study occupies one row of the table (Table 9).

Synopses in clinical section

A well-written synopsis from an individual study report (or the abstract from a publication) should be informative enough to use for the investigator's brochure, but it may on occasion be overly long and require additional summarization. If an investigator requires the full text of a study report, he or she may request it from the sponsor, so there is no need to provide extensive details in the investigator's brochure. Unlike the descriptions of nonclinical studies, information on all clinical studies should be included. An example of a synopsis from a clinical study report may be found in Appendix IX.

Summary of data and guidance for the investigator

The guidance on this section of the investigator's brochure is relatively nonspecific, but has been generally interpreted as a place to begin the product's labeling. The section should, therefore, be structured similarly to a European Summary of Product Characteristics (SmPC) [1] or a Japanese or United States package insert [2–5]. These three documents are used frequently by clinical practitioners because they constitute the primary means by which a drug company conveys product information in Europe and the United States. Japanese package inserts tend to be brief, since Japanese marketing representatives supply much of the information to practicing physicians. The intent of this section is to place all known product information into a package insert format, which is readily recognizable and easily understood.

Use of a competing product's labeling (or a product that has some similarity such as indication or therapeutic class) is helpful in writing this section of the investigator's brochure. Clearly, not all the information from another product's labeling will be available for your investigational product, but it is a starting point. Product labeling is easily available on the Internet. With few exceptions, all of the information in this section (if it exists) is in other sections of the brochure, but the information may require additional summarization or categorization, based on the region in which the information will be submitted.

References

1 European Commission, Enterprise and Industry Directorate-General, A Guideline on Summary of Product Characteristics, October 2005. Revision 1.
2 Guidelines for Package Inserts for Prescription Drugs (Notification No. 606 of PAB dated April 25, 1997).
3 Guidelines for Package Inserts for Prescription Drugs (Notification No. 59 of the Safety Division, PAB dated April 25, 1997).
4 Guidelines for Precautions for Prescription Drugs (Notification No. 607 of PAB dated April 25, 1997).
5 Title 21 – Food and Drugs Chapter I – Food and Drug Administration Department of Health and Human Services Subchapter C – Drugs: General Part 201 Labeling, Subpart B Labeling requirements for prescription drugs and/or insulin, Section 201.56 [71 FR 3986], January 24, 2006.
6 ICH Harmonised Tripartite Guideline E6(R1) Guideline for Good Clinical Practice: Consolidated Guidance, ICH June 1996, http://www/ich.org (Accessed 27 February 2007).

Targeted Regulatory Writing Techniques. Clinical Documents for Drugs and Biologics,
edited by Linda Fossati Wood and MaryAnn Foote
© 2009 Birkhäuser Verlag Basel/Switzerland

Chapter 8.

Investigational medicinal products dossier

Linda Fossati Wood

MedWrite, Inc., Westford, Massachusetts, USA

Introduction

Before human clinical trials can be started in the European Union (EU), the sponsor must request authorization to conduct clinical trials through a submission called a Clinical Trial Authorisation (CTA). This application includes a group of scientific documents called an Investigational Medicinal Products Dossier (IMPD). The EU has provided for two types of IMPDs, a full IMPD and a simplified IMPD, based on whether the product has been described previously in another CTA or a marketing authorization application [1]. This chapter discusses only the full IMPD, since the complexities inherent in determining the type and level of documentation to include in a simplified IMPD are beyond the scope of this book.

Guidance on the structure and content of an IMPD is provided by the European Commission (EC) in ENTR/F2/BL D(2003) CT1 Revision 2, dated October 2005. The IMPD consists of a group of documents, with cross-references to other documents, such as the investigator's brochure, the clinical protocol, or another IMPD. The IMPD has a general structure:

- Quality (chemistry, manufacturing, and controls) data
- Nonclinical pharmacology and toxicology data
- Previous clinical trial and human experience data
- Overall risk and benefit assessment

Because regulatory writers usually only write the text for the clinical sections (previous clinical trial experience and the overall risk and benefit assessment), the chapter focuses on clinical sections. Should the regulatory writer be asked to write the nonclinical material, it is possible to use the same strategies as described for the clinical text.

Previous clinical trial and human experience data

The guideline for the IMPD has a suggested outline for clinical information (Appendix XII). The contents of the outline are similar to those found in the Common Technical Document (CTD) outline for the clinical sections of the CTD in module 2, but the numbering system is different. The IMPD suggested outline is not considered to be exhaustive and does not need to be followed exactly [1].

In general, at the time the CTA is submitted, the investigational product has likely not been used in humans, so it is possible that the section will have no data. In this situation, it is possible to state 'Not applicable.' On occasion, a CTA may be required despite the fact that human testing has already been conducted on the investigational product. It may be that testing took place in a different region of the world, the indication tested was different, or some change was required to the manufacturing methods. In the event that human data are available, there are two general strategies for writing the clinical section of the IMPD.

The first strategy cross-references the investigator's brochure (that should have existing human data included in text and tabular format) and deletes the suggested outline. This strategy is appropriate if the brochure contains sufficient detail for an assessment of safety.

The second strategy cross-references the investigator's brochure but uses sections of the IMPD outline to expand on details that are also summarized in the investigator's brochure to allow the health authority to fully understand safe use of the investigational product. Details in the brochure that are expanded should be placed in the appropriate section of the IMPD outline. These expanded details generally come from clinical study reports, although they may also be found in published literature. Study reports do not need to be submitted.

Overall risk and benefit assessment

This section is intended to be a brief and integrated (ie, including quality information and both nonclinical and clinical research) summary of all risks (safety concerns) and potential benefits to a subject undergoing testing in the proposed clinical trial. The description should include anticipated risks (based on quality, nonclinical, and clinical study data) and methods to be used to attenuate these risks to subjects in the proposed clinical trial (as described in the protocol).

An example illustrates the types of information to consider for the risk and benefit assessment and a proposed associated risk and benefit statement for the fictional drug xerimax:

Quality:
Store xerimax at 2°–4°C, reconstitute with 10 mL normal saline, then rotate the syringe gently until the solution is clear.

Nonclinical:

Chronic administration (28 days) of xerimax to rats and dogs has been associated with hematopoietic suppression (15% reduction in red blood cell counts in rats, 10% reduction in red blood cell counts in dogs) on day 28. Recovery to baseline values occurred by the end of the 2-week recovery period. No other hematologic effects were noted. Clinical observations noted reduced body weight gain in rats. Neither species showed signs of gastrointestinal distress (vomiting [dogs], diarrhea, or reduced appetite).

Clinical:

Design features of the clinical protocol include:

- Inclusion criteria: subjects with hemoglobin concentration ≥12 g/dL (men) and ≥11 g/dL (women); and hematocrit ≥35% (men) and ≥32% (women).
- Body weight will be measured at baseline, weekly during dosing, and at the follow-up visit.
- Clinical laboratory testing for hemoglobin concentration and hematocrit will be performed at baseline, weekly during dosing, and at the follow-up visit.
- Subjects will be removed from the study for substantial reductions in hemoglobin concentration or hematocrit values.
- Subjects will be followed for 1 month after their last dose of xerimax.
- Instructions for storage and reconstitution are included.

Risk and benefit assessment:

Xerimax has generally been well tolerated in nonclinical models, based on laboratory tests and clinical observations. Reduction in red blood cell counts was noted in both species after 28 days of dosing, but levels returned to baseline values by 2 weeks after dosing. Reduced body weight gain in rats was not accompanied by diarrhea or reduced appetite.

The proposed clinical protocol plans to reduce the potential for post-dose anemia by excluding subjects with low hemoglobin concentrations or hematocrit values before enrollment and monitoring values weekly during dosing. Dosing with xerimax will be stopped for any subject with clinically significant reductions in hemoglobin concentration or hematocrit, and the subject will be followed weekly until levels return to baseline values. Body weight will be measured at baseline, weekly during drug administration, and at the follow-up visit.

The protocol includes detailed instructions for reconstitution and administration of xerimax to reduce the potential for intravenously administered particulates. The current formulation has a white, cloudy appearance when first reconstituted with normal saline, and may contain particulates. Gentle rotation of the vial for 1 minute results in a clear solution. Testing has confirmed that the clear solution does not contain particulate matter.

One of the challenges of writing for a large and diverse geographic region such as the EU is attempting to follow a regulation or guideline, while keeping in mind that each of the European member states may have preferences that are not specifically written down. The suggestions described in this chapter represent generalizations and may require modifications for a specific country. These modifications are generally the result of the sponsor's experience with submissions to that country, or negotiations and suggestions from the regulatory body.

References

1   European Commission, Detailed guidance for the request for authorization of a clinical trial on a medicinal product for human use to the competent authorities, notification of substantial amendments and declaration of the end of the trial, ENTR/F2/BL D(2003) CT1 Revision 2. October 2005.

Targeted Regulatory Writing Techniques. Clinical Documents for Drugs and Biologics, edited by Linda Fossati Wood and MaryAnn Foote
© 2009 Birkhäuser Verlag Basel/Switzerland

Chapter 9.

Integrated summaries of safety and efficacy

Jennifer A Fissekis

Rye Brook, New York, USA

Introduction

The Integrated Summary of Safety (ISS) and Integrated Summary of Efficacy (ISE) are separate documents unique to regulatory submission for the United States. They are submitted to the Food and Drug Administration (FDA) in a New Drug Application (NDA) and are not required for European or Japanese submissions. Both the ISS and the ISE are high-level documents and are generally not recommended as a task for the novice writer.

Both the ISS and ISE are integrated documents: They describe the results of more than one clinical trial. Results of all clinical studies performed on the investigational product are generally combined into one database (called pooling), so results are statistically summarized as a whole. This pooling results in a database that is much larger than the individual study databases and therefore has a better chance of detecting statistically significant differences between treatment groups. Statistical differences are useful not only for detecting efficacy but also for finding problems with safety that might have been missed in a small dataset. A case of hepatic failure here and there in a few individual studies stands out more clearly when clustered together in a pooled database. The function of this overall (or pooled) evaluation using integrated analyses of all data obtained is to support use for the indication being investigated.

What the ISS and ISE comprise

In July 1988, the FDA published guidance on the Format and Content of the Clinical and Statistical Sections of an Application (Clin-Stat Guidance) in which the contents of the ISS and ISE were described [1]. In 2001, the broad outline of the ISS and ISE in this guidance was partially updated by the ICH guidance M4 Common Technical Document for the Registration of Pharmaceuticals for Human Use [2]. These two guidances currently present the regulatory view of requirements for the content of the ISS and ISE for an NDA submission. They provide the reviewer with an overall summary that is a critical evaluation of the information generated by the sponsor to determine whether the drug is safe and effective in its proposed use(s) and whether the benefits of the drug outweigh the risks.

Table 1 presents a list of critical elements for the ISS and Table 2 a list of critical elements for the ISE. Both the ISS and ISE should be placed in Section 5.3.5.3 of the Common Technical Document (CTD) [3].

Table 1. Critical elements of the ISS

- An evaluation of the overall summaries and statistical analyses of the combined safety data from the various clinical studies conducted for licensure.
- An examination of the incidence of adverse effects in various subgroups of the patient/subject database. The effect of sex, ethnicity, or age (adolescent/young adult versus older adult), if any. The data from pediatric trials among children aged less than 13 or 14 years are usually not combined with the data from older age groups, recognizing that children are not 'small adults' either in regard to the appropriate dosage or to the anticipated adverse effects.
- The effect on safety and efficacy of concomitant medications taken by study subjects.
- Assurance that the proposed dose is appropriate regardless of age, and renal and liver function
- If the drug product is for a chronic condition, assurance that data support its long-term use.
- Identification of concerns with use of the test article with other medications that the patients in the target population may be taking.
- Assurance that the presentation of the data supports the concept that the benefits of the drug product outweigh the risks associated with its use.

Table 2. Critical elements of the ISE

- An evaluation of the overall summaries and statistical analyses of the combined efficacy data from the various clinical studies conducted for each claimed indication for licensure.
- An examination of the evidence supporting the dosage and administration interval in various subgroups of the patient/subject database. The effect of sex, ethnicity, or age (adolescent/young adult versus older adult), if any. The data from pediatric trials among children less than 13 or 14 years of age are not normally combined with the data from older age groups, recognizing that children are not 'small adults' either in regard to the appropriate dosage or to the anticipated adverse effects.
- The effect on efficacy of concomitant medications taken by participants in the studies.
- Is the proposed dose appropriate regardless of age, and kidney and liver function?
- If the drug product is for a chronic condition, are there data supporting its long-term use?
- Is there any evidence that there is a therapeutically active metabolite that contributes to the efficacy or adverse effect profile?
- Does the presentation support the concept that the benefits of the drug product outweigh the risks associated with its use?

The brief paragraphs of the guideline show that a large amount work is required, which may determine the success of the application. The ISS and ISE are not mere summaries and recapitulations of the information presented in the studies but rather critical assessments of the integrated data. The ISE is the overall document that must serve not only as a persuasive argument for approval to market but will also provide the information that guides the safe and effective use of the drug product by physicians and patients. The results of the analyses and critical evaluation of the efficacy presented in the ISE will determine the specific indication for which the drug product is approved (licensed) and will affect what is required to be contained in the patient information placed in the package for marketing. The labeling will establish and limit the approved use of the drug for compensation by third-party payers (insurance companies) and may also affect whether a drug is considered for inclusion on formulary listings, such as those maintained by health maintenance organizations (HMOs) and the Veterans Administration. It is important that as this document is written, the objectives of the sponsor be supported by the data presentations.

The prime requirement directing the preparation of an effective ISS or ISE is to establish, up front, the corporate message, the marketing goal, the therapeutic niche the product is to fulfill, and how the product compares with other products already in the market (if any). These products may in fact have been used in the clinical studies as active therapeutic comparators. Presentation of the data must support these objectives and validate the efficacy of the proposed new therapeutic entity. If planning has been effective and writing is directed towards the goals, there are sentences in the conclusion and summary subsections of the nonclinical and clinical study reports that can be lifted verbatim into the ISE to create a cohesive document.

The writer's role in the ISS and ISE
The writer is often the primary manager in the process of completing this section of the NDA. As the manager of the project, it is the job of the regulatory medical writer to obtain agreements on reviewer responsibilities and timelines from the contributors. At minimum, the contributors/reviewers would be team members from nonclinical toxicology, the pharmacology and pharmacokinetics departments, the clinical study group, statisticians and statistical programmers, and a representative of regulatory affairs. Both the ISS and ISE are of necessity documents that are written after all the study reports have been completed. If any of these reports have been completed later than originally planned, as almost invariably happens, the project timelines may already be impacted. Only rarely will the failure to complete study reports on time lead to an extension in the date that the ISS or ISE is due.

The design of tables and their content should have been agreed on while the clinical studies were in progress. Planning requires that time must be allowed for the usually neglected steps for statistical programmers to write the codes for the production of the tables required and to validate that the code correctly extracts the data. Test tables should be produced for reviewer approval and to show the final format on a page. This process should help make high-level reviewers aware that last-minute requests for tables will delay completion of the project and compromise meeting submission deadlines. These processes are not merely 'nice to have,' they are required, by regulation of the design and execution of statistical analyses, to have validated analyses whose results are accepted.

Achieving consensus from various departments contributing information for the ISS and ISE is difficult but essential to the process. Writing these important summary documents is a cross-functional exercise in obtaining information from many people and gaining consensus on data presentation. It is critical to the success in the writer's role to recognize the importance of accepting that each contributor has in the past had experience with presenting information to the Agency and has valuable insight into the thinking of the reviewers in their discipline. Negotiation is a necessary process to obtain a product (the written document) that will meet corporate goals. Some of the pride of authorship will be missing since the writer's job will be to synthesize the input from other contributors, each of whom thinks his/her contribution is of prime importance. They will not always appreciate that the sterling information they contribute, gained by meticulous and often impressive scientific endeavor, is of less importance in the overall summary than their scientific insight.

All the data are presented in support of the objectives for the product, which does not mean that the data are twisted or spun, only that a clear message is supported consistently throughout the application. The message components arise from the data produced in the clinical studies and supported by the nonclinical study information. If planning has been effective and writing is directed towards the goals, sentences in the conclusion and summary subsections of the nonclinical and clinical study reports can be used verbatim in the ISS and ISE to create a cohesive document. Such a technique allows hyperlinking within the electronic document and permits the reviewer to go directly to the detailed report that supports the statement. Repeating the wording verbatim from other sections of the submission allows for quick recognition of where in the report the information is presented. It has always to be borne in mind that the reviewer does not have the intimate knowledge of the information gained during the development of the drug product, and needs to be helped to reach the same conclusions that are almost self evident in the minds of those who have spent years bringing the drug product to the point of submitting an NDA. This goal requires that the by now intuitive knowledge of the developers be presented in a coherent sequence to allow the reviewer to see how conclusions were reached. As a practical matter, the repetition of phrasing also makes the work of those responsible for putting in the hyperlinks and validating their accuracy quick and easy.

Clearly, the preparation of the ISS or ISE is not for the new medical writer. These are documents that such a writer could edit for technical format and grammatical accuracy but not write without considerable mentoring. In editing, however, attention must be paid to the overall message and intended use of the document. To edit effectively requires that the writer/editor appreciate the purposes of the document and how it is generated with data from many areas of expertise, not only from the overall tabulations of reported clinical adverse events and clinical chemistries, but also with

---

Side bar: Historical development of the regulations for safety and efficacy

Requirements for evidence of safety and efficacy of new drugs are the result of a number of unfortunate occurrences such as the marketing of an untested formulation of sulfanilamide, a new wonder drug, in 1937 [4]. The mislabeled elixir contained diethylene glycol, a substance that causes renal toxicity and resulted in more than 100 deaths, many of them children being treated for infections. A second notable milestone in the development of drug regulations during the 1960s was prompted by the drug Kevadon, a brand name for thalidomide, which was approved for marketing outside the United States. The medical officer at FDA, Dr Frances Kelsey, refused to allow the approval of the Kevadon New Drug Application because insufficient safety data were presented. In 1962, the devastating effects of this drug on bone development (osteogenesis) in the fetus became evident when phocomelia and other deformities due to in utero exposure to the drug were observed in countries that allowed commercial distribution of Kevadon. Dr Kelsey received the President's Distinguished Federal Civilian Service Award in 1962 for her efforts to prevent the licensure of the drug in the United States [4].

---

references to the nonclinical safety data. With experience in preparing clinical study reports and in participating in the team effort that leads to the overall assessment, the writer can become a contributing member of the team. In all cases, it will need significant diplomatic skills, and a willingness to listen to the needs described by all scientific areas. This will enable the production of one of the critical pieces of a submission that must serve not only for approval to market but guide the use of the drug product by physicians and patients.

Several points of concern may have been discussed with the FDA during the development of the drug product, and these concerns should be addressed in the ISS or ISE. Any special concerns that the FDA has expressed should be specifically described and the answers presented. It will be necessary for the writer to check with the regulatory affairs team member to obtain copies of any letters in which questions were asked or addressed. Specific adverse effects captured as a result of the design of the protocol are of primary interest since these are considered a priori to be possible indicators of problems.

References

1 Guideline for the Format and Content of the Clinical and Statistical Sections of an Application http://www.fda.gov/cder/guidance/statnda.pdf July 1988. (Accessed 24 March 2008)
2 M4, The Common Technical Document, Safety at http://www.ich.org/cache/compo/276-254-1.html

3   FDA Centers for Drug and Biologics Evaluation and Research (CDER and CBER). Guidance for
    Industry: Integrated Summaries of Effectiveness and Safety: Location Within the Common Technical
    Document. Draft, June 2007. Available at http://www.fda.gov/cder/guidance/7621dft.htm (Accessed 17
    April 2008)
4   www.FDA.gov home page, FDA History, Milestones in FDA History Backgrounder: http://www.fda.
    gov/opacom/backgrounders/miles.html (Accessed 24 March 2008)

Targeted Regulatory Writing Techniques. Clinical Documents for Drugs and Biologics,
edited by Linda Fossati Wood and MaryAnn Foote
© 2009 Birkhäuser Verlag Basel/Switzerland

Chapter 10.

Informed consent forms

Jennifer A Fissekis

Rye Brook, New York, USA

Introduction

Informed consent is obtained from a possible participant in a clinical trial primarily to protect the rights, safety, well being, and interests of those participating in the trial. Written consent has been developed to avoid coercive or deceptive recruitment and the use of unethical enrolling practices.

Of necessity and by regulatory requirements, consent is usually obtained before any procedure involved in the trial is undertaken. Under certain defined or exceptional circumstances, informed consent may be given by someone other than the participant. In all cases, however, the process must be approved by an Institutional Review Board (IRB, in the United States and Japan) or Independent Ethics Committee (IEC, in the European Union). A brief overview of the historical development of the current requirements for obtaining informed consent and the content of the form is provided in this chapter. The required elements of informed consent, as detailed in the International Conference on Harmonisation (ICH) of Technical Requirements for Registration of Pharmaceuticals for Human Use Guidelines for Industry and the United States Food and Drug Administration (FDA) regulations, are described. Two regulatory documents specify the requirements for informed consent: the Guidance for Industry E6 Good Clinical Practice: Consolidated Guidance, section 4.8 [1] and CFR Title 21 Sections 50.20 and 50.25 [2]. The documentation of signature of written informed consent is described in 21CFR section 50.27. The investigator should adhere to Good Clinical Practices (GCP) requirements and to the ethical principals that have their origin in the World Medical Association (WMA) Declaration of Helsinki.

Historical development of regulations governing informed consent

The history of informed consent as it is required today derives from not only ethics as expressed in the best practice of medicine but also from the ethical failures of some investigators [3]. Most writers in the field of regulatory medicine have at least passing knowledge of the abuse of prisoners at the Tuskegee Institute, where men with known syphilitic infections were left without antibiotic therapy to observe the progress of untreated disease over many years. This incident happened at a time when an effective treatment of the disease was available. Another incident in the early 1960s illustrated the need for general control over the recruitment of subjects for clinical trials. In an effort to determine the source of the ability of cancer cells to escape destruction by the body's immune responses, live HeLa (breast cancer) cells were injected into elderly, and in some cases immunocompromised or senile patients, to see if their immune systems would destroy the cells or if they would grow into a tumor in a subject with compromised immune responses [3]. Oral 'informed consent' had been obtained from the subjects or in some cases given by the operator of a nursing home for those incapable of understanding what was being explained to them. The subjects were told that their 'resistance' was being tested. Lastly, the memory of abuses of prisoners during World War II resulted in the Declaration of Helsinki, first adopted in 1964 and amended five times. This document embodies ethical principles to provide guidance to physicians and other participants in medical research involving human subjects [4].

Article 8 of the Declaration of Helsinki addresses specifically the responsibilities one has concerning the rights and protection of vulnerable populations: "Medical research is subject to ethical standards that promote respect for all human beings and protect their health and rights. Some research populations are vulnerable and need special protection. The particular needs of the economically and medically disadvantaged must be recognized. Special attention is also required for those who cannot give or refuse consent for themselves, for those who may be subject to giving consent under duress, for those who will not benefit personally from the research and for those for whom the research is combined with care."

The Kefauver-Harris Amendments signed into law on 10 October 1962 also instituted stricter agency control over drug trials including a requirement that patients involved must give their written informed consent [5].

Process of obtaining informed consent

Important items to be considered in obtaining informed consent are provided in Table 1. To recruit participants for a clinical trial, an investigator who is considered likely to have the patient (or subject) population that is required for the particular trial is approached to participate. The investigator must present the proposed clinical protocol and supporting documents to the IRB or IEC and receive their approval for the study to be conducted before speaking to potential participants. When the IRB/IEC approval is given, those identified as potential candidates to be included in the study, that is persons who meet the inclusion and exclusion criteria of the protocol, can be approached using information leaflets, flyers, or advertisements or by telephone contact. Before enrollment of a subject or patient, and before any study-related procedures are performed, voluntary written, study-specific, informed consent must be obtained.

Table 1. What is required to obtain informed consent

- A consent document
- The process of ensuring comprehension of the document's contents
- Ensuring that consent is obtained from the participant or his or her legally authorized representative in the case of a minor or a subject with mental or physical impairment that prevents them from giving consent
- The authorized site personnel (physician, nurse, coordinator) must be personally involved and sign and date the consent form
- Signatures and dates must be entered by the person giving consent and by the person obtaining the consent each in his or her own handwriting
- Source documentation must be archived and available for inspection

---

Leaflets and informed consent forms must be provided in a language that the subject (or parent or guardian of a minor child) can read, and the discussion of the content must be conducted by the investigator or an authorized (medically qualified) designee. Sufficient time must be allowed for the person giving consent to consider the information and think about the implications of participation in the study. The investigator or the authorized (medically qualified) designee obtaining the informed consent must also sign the form. Obtaining consent can be done on the same day as the information and consent form is given to the potential participant, or may be obtained subsequent to the provision of the documents to the potential participant. In no case can the form be signed by the person obtaining the consent before the subject has signed and dated the form in their own handwriting, or that this has been done for them by a witness in the case where the consent giver is illiterate. Regulations vary in different countries. In some countries, only one parent needs to sign for a minor child, in other countries, both parents need to agree to the child's participation in the study. The physician or regulatory designee in the organization conducting the study is responsible for providing this information to whomever is preparing the informed consent form so that provision can be made for the correct signatures to be obtained. No national ethical, legal, or regulatory requirement should be allowed to reduce or eliminate any of the protections for human subjects set forth in the Declaration of Helsinki.

Typically, a letter is drafted by physicians or medical monitors from the sponsor developing a drug, and it is printed on the letterhead of the participating investigator, who will explain the contents to the potential participant. The letter is required to present information in simple language, translated if necessary from English (in the United States of America or Great Britain, or the originator's language in other countries) into the language of the area from which the participants are to be recruited. At the end of the letter, a form, preferably a single page, is included that will be signed and dated by the participant (or parent/guardian of a minor child) and by the person who obtains the informed consent. On this page, statements that capture the main points of the informed consent appear, and each of these statements should be initialed by the participant to acknowledge receipt and understanding of the information. If the person giving consent is not able to read, then a witness to the oral presentation of the information and explanation must also sign and date the informed consent.

Elements of the informed consent form

An example of an informed consent form may be found in Appendix XIII. To meet the regulatory requirements for an informed consent form, several elements should be included (Table 2). An example of the language to be used in a paragraph to meet the needs of element 1 of the table could be:

> <<*Insert the name of the practice or facility at which the study will be conducted*>> is carrying out a research study of a new <<*drug/vaccine/procedure/medical device*>> to protect against <<*insert the disease condition under investigation in simple layman's terms, if possible*>>. We would like to ask you to take part in this study. This letter explains what the study is about and what would be involved if you decided to take part in it. This letter has been <<*given/posted/mailed*>> to you by <<*insert contact details*>>. The study is being conducted with <<*name of sponsor organization*>>, referred to in this letter as the 'Sponsor,' who will be paying for the study.

An example of the language to be used in a paragraph to meet the needs of element 2 of the table could be:

> If you agree to take part in the study, you will be treated (or receive study drugs) each month for 12 months from when you were enrolled and were randomly assigned to a study group. We will follow-up by telephone 6 months after your last visit to see if any new effects have occurred since the end of the study and to check on your general health. There will be about <<*number*>> subjects enrolled into this study in all groups. We expect the study to be completed in about <<*number*>> months. [**Or give a specific date.**]

Table 2. Required elements of informed consent document

No.	Element	ICH GCP 4.8.10	FDA CFR 50.25
1	The study involves research, and the purpose of research	a, b	a1
2	Duration of participation and approximate number of participants (subjects or patients)	s. t	a1, b6
3	Description of procedures and the probability of random assignment	c, d	a1
4	Identification of experimental aspects of the study	f	a1
5	Description of foreseeable risks or discomforts, use of an approved adverse event profile	g	a2
6	Benefits to the participant or others	h	a3
7	Alternative procedures or courses of treatment and important benefits and risks	i	a4
8	Participant's responsibilities	e	
9	Extent of confidentiality of records, and personal identification information	o	a5
10	Information that the regulatory authorities, FDA, IRB/IEC and sponsor's representatives may inspect records	n	a5
11	Compensation and treatments in case of injury	j	a6
12	Contacts (name, telephone number) for study, rights as a research subject, research-related injury, treatment	g	a6, a7
13	Statement that participation is voluntary	m	a8
14	Refusal to participate or withdraw without loss of benefits to which the participant is otherwise entitled	m	a8
15	Participant to receive a signed and dated copy of the informed consent	4.8.11	50.27
16	Unforeseeable risks to study participant, embryo, or fetus; nursing infant, pregnant woman, men (when appropriate)		b11
17	Investigator withdrawal of a participant without consent of participant		b2
18	Additional costs, if any, to the participant resulting from participation in the study	l	b3
19	Payment (include payment schedule, prorating if appropriate)	k	
20	Foreseeable circumstances and/or reasons why participant may end involvement with the study	r	b2, b4
21	Termination procedures		b4
22	Significant new findings and willingness to continue participation	4.8.2; 4.8.10p	b5
23	Premature termination or suspension of a trial	4.12	
24	Signature and date lines	4.8.8	50.27

FDA CFR, Food and Drug Administration Code of Federal Regulations; ICH GCP, International Conference on Harmonisation Good Clinical Practices; IEC, Independent Ethics Committee

In each case and for each element, the precise wording will depend on the specific study protocol and must reflect the conditions of the study as set out in the protocol. Similar paragraphs will present the information derived for each element in simple language and terms understandable by the community identified as potential study subjects, that is, subjects who meet the study population requirements of the protocol or the subject's legally acceptable representative (parent or guardian for a child, spouse, or health proxy holder for a mentally or physically incapacitated adult). Essentially, the paragraphs are answers to the "Who, Why, What, Where, When, How" questions that could be asked and they provide the information detailed in the list of essential elements of informed consent. The language used in both the written and oral presentation of information about the trial should be as nontechnical as possible and should be understandable to the subject or his/her representative, and the impartial witness.

It is recommended that the approximate education level of the participants be considered to be no more than United States grade 8 (elementary school) and that an appropriate readability test be used to achieve this goal. In all cases, after the letter has been read by the potential participant, an opportunity for them to discuss and clarify any of the information with the investigator or a designated study nurse who is capable of answering all queries should be provided. All questions about the trial should be answered to the satisfaction of the subject or the subject's legally acceptable representative.

An opportunity should be given to the participant, or parent of a minor child, to discuss the letter with their family, or other advisor as they choose, before giving consent. A statement to this effect should be included in the letter.

Provisions are made in the guidances and in the Declaration of Helsinki [4] for subjects to be enrolled in a clinical trial under conditions where they cannot give informed consent (21 CFR 50.24, subpart B; ICH E6(R1) section 4.8.12 to 4.8.15) [1, 2]. Such circumstances could be emergency situations when consent is not possible from the subject. These groups should not be included in research unless the research is necessary to promote the health of the population represented and this research cannot instead be performed on legally competent persons. Under such conditions, consent should be sought from the subject's legally acceptable representative or surrogate, if present. If the subject is unable to give consent and his or her legally acceptable representative is not available, the subject can be enrolled if measures to do so are described in the protocol, with documented approval of favorable action of the IRB/IEC, and measures are described to protect the rights, safety, and well being of the subject.

Some situations exist in which surrogate consent is not obtainable, or when consent would be sought later when the subject was able to give consent, or when the subject's representative could later be informed and consent to continue obtained. One such situation would be the testing of a psychoactive drug for extreme mental disturbance or psychotic episode in a subject having the disease or condition for which the investigational product is intended as therapy. If the trial drug is effective, consent to continue the trial should be obtained from the subject as soon as possible.

Nontherapeutic trials, which are trials in which there is no anticipated direct clinical benefit to the subject, should be conducted in subjects who personally give consent and sign and date the written informed consent form. Provision is made in the regulations for non-therapeutic trials to be conducted in subjects with the consent of a legally acceptable surrogate. Certain conditions need to be fulfilled (Table 3).

Table 3. Provision for nontherapeutic trials in subjects with the consent of surrogate

---

- The objectives of the trial cannot be met by means of a trial in subjects who can give informed consent personally
- The foreseeable risks to the subjects are low
- The negative impact on the subject's well being is minimized and low
- The trial is not prohibited by law
- The approval/favorable opinion of the IRB/IEC is expressly sought on the inclusion of such subjects, and the written approval/favorable opinion covers this aspect

---

In cases where investigational treatment is provided without written informed consent, it is critical that complete, accurate, legible reports and records be maintained in a timely manner. Subjects should be closely monitored and withdrawn from the study if they appear to be unduly distressed.

References

1   ICH Harmonised Tripartite Guideline E6(R1) Guideline for Good Clinical Practice: Consolidated Guidance, ICH June 1996, http://www/ich.org (Accessed 23 March 2008); ICH Guidelines, Efficacy topics: http://www.ich.org/LOB/media/MEDIA482.pdf (Accessed 23 March 2008).
2   FDA Code of Federal Regulations. 21CFR 50.20 and 50.23 and 50.24 and 50.25 at http://www.fda.gov. Database updated April 1, 2007 (Accessed 23 March 2008).
3   FDA History. http://www.fda.gov/oc/history/default.htm (Accessed 23 March 2008)
4   The current version of the Declaration of Helsinki, 2004, is the only official one; all previous versions should be used or cited as historical references only. Available at The World Medical Association, http://www.wma.net/e/policy/b3.htm. (Accessed 23 March 2008).
5   FDA CDER History. http://www.fda.gov/cder/about/history/page32.htm (Accessed 23 March 2008).

Regulatory submissions

Targeted Regulatory Writing Techniques. Clinical Documents for Drugs and Biologics,
edited by Linda Fossati Wood and MaryAnn Foote
© 2009 Birkhäuser Verlag Basel/Switzerland

Chapter 11.

Global submissions: The common technical document

Peggy Boe

Image Solutions, Inc., Wilmington, North Carolina, USA

Background

Historical perspective

The process of discovering and developing a new product can easily take 10–15 years and can cost a sponsor hundreds of millions of dollars. Yet only about 1 of 10 000 potential products is deemed to be safe and effective and therefore eligible for approval [1]. In addition, with the widespread use of improved computer technology, the international public has become increasingly aware of the risks associated with using drugs and biologics. Sponsors and regulatory agencies both are being held accountable for approved marketing of products that are found to have life-threatening consequences once they are used by a large population. As a result of public pressure for improved safety and of industry pressure for regulatory agencies to streamline the approval process, many initiatives have arisen that have driven the need to develop the ability to share safety information in a consistent way on an international level [2].

Table 1 outlines some of the historical events related to the advent of product approval requirements and safety initiatives at the US Food and Drug Administration (FDA), as an example of one region's attempts to address product safety and regulatory agency efficiencies [3].

Table 1. Milestones in US food and drug law history [3]

Year	Event
1820	• First US Pharmacopeia, a compendium of standard drugs for the US
1862	• First Bureau of Chemistry, the predecessor of the FDA
1902	• Biologics Control Act passed to ensure safety of serums, vaccines, and other products to be used in humans
1906	• Original Food and Drugs Act passed to prohibit interstate commerce in misbranded and adulterated food and drugs
1914	• Harrison Narcotic Act required prescriptions for narcotics and increased record-keeping by physicians and pharmacists
1937	• Elixir of sulfanilamide killed 107 people
1938	• Federal Food, Drug, and Cosmetic (FDC) Act of 1938 began a new system of drug regulation requiring new drugs to be shown safe before marketing
1949	• FDA published first Guidance to Industry
1950	• Court ruled that the directions for use on a drug label must include the purpose for which the drug is offered
1962	• Thalidomide found to cause birth defects; public demanded stronger drug regulation • Kefauver-Harris Drug Amendments, passed to ensure greater drug safety and to require drug manufacturers to prove efficacy before marketing • Consumer Bill of Rights included the public's right to safety, to be informed, to choose, and to be heard
1970	• FDA required the first patient package insert to contain information for the patient on a product's risks and benefits
1972	• Over-the-counter (OTC) drug review began
1979	• National Institutes of Health (NIH) issued the Belmont Report, including statements and guidelines on ethical principles associated with the conduct of research with human subjects
1983	• Orphan Drug Act promoted research on drugs needed to treat rare diseases
1988	• CDER issued guidance on the contents of the clinical and statistical sections of a new drug application
1992	• Prescription Drug User Fee Act required manufacturers to pay fees to the FDA for processing applications, to be used to fund increasing staff and making improvements in review processes and approval timelines
1993	• MedWatch launched, a reporting system for adverse reactions
1997	• FDA Modernization Act renewed the PDUFA Act of 1992 and included measures to accelerate review of devices and to regulate advertising of unapproved uses of approved drugs and devices
1998	• Pediatric Rule required manufacturers to conduct studies of safety and efficacy in children in select drugs and biologics
1999	• CDER and CBER issued a guidance on general considerations for electronic submissions to the agency • ClinicalTrials.gov founded to inform the public of current clinical research activities
2000	• CTD Guidelines finalized by ICH

Table 1 (continued)

Year	Event
2002	• Current Good Manufacturing Practice (cGMP) initiative applied consistent quality standards to products and manufacturing processes
2003	• CDER and CBER issued additional guidance on general considerations for electronic submissions to the Agency • CTD mandatory for the EU and Japan, and highly recommended for the US, after July 1
2005	• Formation of the Drug Safety Board, including members of FDA, NIH, and the Veterans Administration (VA), to improve communication of safety issues to the public • CDER and CBER issued a guidance for submitting product labels electronically
2006	• CDER withdrew previous guidance for submitting NDAs, ANDAs, and Annual Reports in the previous electronic format • CDER and CBER issued new guidance on using the eCTD specifications for human pharmaceutical product applications, applicable to Abbreviated New Drug Applications (ANDAs), Biologic License Applications (BLAs), Investigational New Drug Applications (INDs), New Drug Applications (NDAs), master files, advertising material, and promotional labeling
2007	• Food and Drug Administration Amendments Act (FDAAA) of 2007 reauthorized and expanded PDUFA
2008	• CDER posted notice that, as of 01 January 2008, CDER will only accept NDA, IND, ANDA, BLA, Annual Report and Drug Master File submissions as all paper or as an eCTD, and sponsors must request a waiver to submit in any electronic format other than the eCTD

Region-specific submission content

Sponsors seeking product approvals globally must apply to numerous agencies, each with country-specific regulations. Until recently, submitting product applications to the three main regions – European Union (EU), Japan, and the United States – required submitting three completely different applications of varying complexity. The burden of understanding the regional requirements and of creating the different components typically fell on a sponsor's regulatory affairs personnel, and it was almost impossible to even consider submitting to more than one region at a time. Revising the original dossier to meet another region's standards could take months, greatly increasing the cost and time spent trying to get a product approved for global marketing. The single most important initiative to improve all of the abovementioned concerns (ie, difficulty marketing important products globally because of regional differences, inefficient review processes that delay getting products to market, and insufficient sharing of safety information) was that of moving from region-specific submission content and format to the harmonized dossier known as the Common Technical Document (CTD).

The content required for a New Drug Application (NDA) in the United States, a Marketing Authorisation Application (MAA) in Europe, and a Japanese Gaiyo differed significantly with regard to document organization and submission of summaries and tables. For example, an MAA submission required tabulated summaries and an expert report to be written by an unbiased clinical expert, with the expectation that the report would candidly point out the product's risks and benefits. In the United States NDA, no such expert report was required, but the NDA did require written summaries and reports of pooled data for safety and efficacy to help identify trends in a larger population. As a result, depending on where the original dossier was submitted, individual documents had to be revised or replacement documents created.

To address these concerns, the International Conference on Harmonisation (ICH), which was formed in 1990 with representatives from the regulatory agencies and industry associations of Europe, Japan, and the United States, decided to develop a uniform (common) technical document (ie, the CTD) organization format

Table 2. Global regulatory authorities and regulatory initiatives for drugs and biologics [5]

Geographic region	Regulatory authority
European Union (EU)	European Medicines Agency (EMEA)
Japan	Ministry of Health, Labor, and Welfare (MHLW) Pharmaceuticals and Medical Devices Agency (PMDA)
United States	Food and Drug Administration (FDA) • Center for Drug Evaluation and Research (CDER) • Center for Biologics Evaluation and Research (CBER)

Global initiative organizations	Members
International Conference on Harmonisation	Includes regulators and industry members from 6 founding parties: • European Commission – European Union (EU) (includes the European Medicines Agency (EMEA) and the Committee for Medicinal Products for Human Use (CHMP) • European Federation of Pharmaceutical Industries and Associations (EFPIA) • Ministry of Health, Labor, and Welfare (MHLW) (includes the Pharmaceuticals and Medical Devices Agency (PMDA) • Japanese Pharmaceutical Manufacturers Association (JPMA) • United States Food and Drug Administration (FDA) • Pharmaceutical Research and Manufacturers of America (PhRMA)
	ICH observers who act as links with non-ICH countries: • The World Health Organization (WHO) • The European Free Trade Association (EFTA) • Health Canada
International Federation of Pharmaceutical Manufacturers and Associations (IFPMA)	National industry associations and companies from developed and developing countries

for the dossiers that would be acceptable in all three of the aforementioned ICH regions. Regional differences also included variations in whether paper or electronic submissions were acceptable. Thus, the ICH addressed both the differences in content and the need for a standardized technology that could be accessed and used successfully on a global basis [4]. The resulting ICH M4 Guidelines provided the uniform structure for content, and the ICH M2 Guideline addressed the specifications for the standardized technology [5–9]. Table 2 outlines the major ICH regional authorities and initiative groups [10].

Reaching a consensus for a harmonized global submission structure

As outlined in Table 2, the ICH Committee includes experts from industry and regulatory agencies who meet to create global guidelines based on their areas of expertise. The expert working groups (EWGs) for the CTD included technical experts for the global electronic submission considerations and representatives for safety (nonclinical), quality (chemistry, manufacturing, and controls), and efficacy (clinical) content; the CTD guidelines are thus placed under the category of multidisciplinary. (The ICH Guidelines are all known by an alpha-numeric system: 'S' for nonclinical topics, 'Q' for quality, 'E' for efficacy, and 'M' for multidisciplinary). The CTD guidelines are M4 for all contents and M2 for the technical specifications [4]. All ICH guidelines undergo the same process steps to reach finalization and are overseen by the ICH Steering Committee, which is supported by ICH Coordinators and the ICH Secretariat [1, 7, 11]:

- Step 1: Create an EWG for a new topic and gain expert consensus.
- Step 2: Obtain a signed agreement from the ICH party members that a consensus has been reached and that the draft guideline should proceed to the next step.
- Step 3: Publish the draft guideline for regulatory consultations and discussion (issued as a Committee for Medicinal Products for Human Use (CHMP) Guideline in the EU, translated and issued by the Ministry of Health Labour and Welfare (MHLW) in Japan, and published as a draft guidance in the Federal Register in the United States).
- Step 4: Adopt the ICH Guideline when sufficient scientific consensus is reached.
- Step 5: Implement the ICH Guideline according to national and regional procedures.

ICH committees focus on harmonizing the technical requirements for new drug products; thus the CTD guidelines are specific to new pharmaceuticals, including biotechnology-derived products.

The CTD as a structure for submission content

Goals and objectives of the CTD

The goals and objectives of the ICH M4 EWG included ways to address the concerns of sponsors who wished to submit marketing applications to different regions. It was believed that providing a common format for global submissions could significantly decrease the time and resources that sponsors needed to compile and update submissions and that agency reviewers needed to assess them. From both the sponsor's and reviewer's perspectives, knowing where to place certain information and where to find that information, respectively, should eliminate the guesswork and variation that occurred previously between regions and even between sponsors in the same region. The ultimate goal of the CTD was to enable sponsors to create one dossier that would be acceptable to all ICH regions. To achieve that, the guidelines caution sponsors not to deviate from the overall organization of the CTD, although the content of individual documents should be modified as needed for adequate presentation of the submission content. The guidelines also make it clear that they are not to be considered as guidance on what a sponsor needs to do to develop a product and meet local regulations. Figure 1 provides a diagram of the CTD organization, as presented in the ICH guidelines [5]. This chapter describes how to write Modules 2.5 (clinical overview), 2.7 (clinical summary), and 5 (clinical data) and describes placement of the investigator's brochure in Module 1.

The CTD has provided a means for sponsors to create a registration dossier that is much more harmonized than ever before; however, the CTD in reality is a starting point for unified content and structure, but it is certainly not acceptable as it is in all regions. Although the structure of Modules 2–5 should be the same for all regions, some of the outlined contents may not be required, depending on the region and also depending on the type of submission. The guidelines have undergone several revisions, and questions continue to arise as sponsors become better acquainted with the structure and discover individual variations that are not covered by the guideline. As a result, ICH has issued question-and-answer documents to respond to some of the more frequently asked questions. In addition to working as a team under close collaboration with clinical and regulatory affairs experts, anyone who is assigned to write documents intended for submission as a CTD should prepare by reading all of the guidelines, question-and-answer documents, and local agency regulations and guidance associated with the document to be written to ensure that the content is sufficiently presented and provided in the correct location [2].

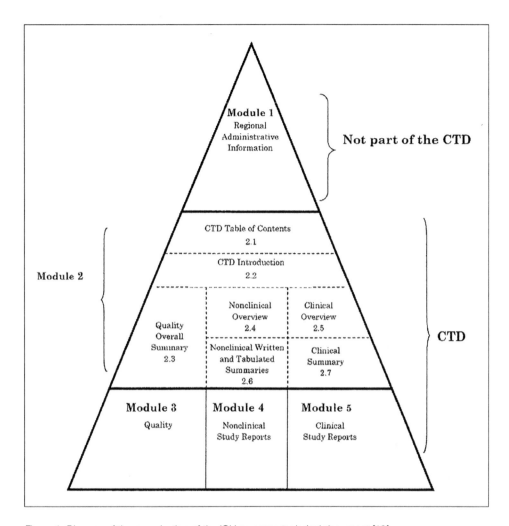

Figure 1. Diagram of the organization of the ICH common technical document [12]

Writing clinical CTD documents – The where, how, and why of it

CTD granularity

Part of what makes the CTD structure unique compared with previous submission formats is what is known as document granularity. Items that used to be submitted as one document are now divided into multiple granular components (multiple electronic files) that are submitted as individual documents. The concept of increased granularity in the CTD was introduced for several reasons, primarily to provide sponsors maximum benefits when the dossier is submitted as an electronic CTD (eCTD). In simplified terms, the granularity that is specified in the CTD divides a document into logical, related components [4].

Module 1: Regional content

The ICH guideline titles Module 1 as 'Administrative Information and Prescribing Information' and indicates that regional regulatory authorities can define both the required content and the structure for providing that content [12]. Therefore, Module 1 is not considered to be part of the common technical document, and the ICH does not include any guidelines for what to include in Module 1, other than an overall submission table of contents (paper submissions only). Module 1 specifics may change occasionally since they are governed by regional authorities; therefore, sponsors should search the various regional agency websites for current and relevant information and forms for Module 1. In US submissions, Module 1 includes the investigator's brochure in Section 1.14.4.1.

Module 2 contents

Module 2 contains various summaries of the collective research reports and data that are provided in the basis of the submission, found in Module 3 (quality) [6], Module 4 (nonclinical) [7], and Module 5 (clinical) [8]. The contents of the Module 2 clinical summaries and reports are described in Table 3.

Table 3. The clinical contents in Module 2

• Module 2.5: Clinical overview
• Module 2.7: Clinical summaries, including:
   • Module 2.7.1: Summary of biopharmaceutical studies and associated analytical methods
   • Module 2.7.2: Summary of clinical pharmacology studies
   • Module 2.7.3: Summary of clinical efficacy
   • Module 2.7.4: Summary of clinical safety
   • Module 2.7.5: References
   • Module 2.7.6: Synopses of individual studies

The clinical overview

The clinical overview (Module 2.5) is intended to be a high-level discussion providing an expert opinion of the overall research and development strategy and results and a discussion of safety and efficacy findings, with an emphasis on the clinical risks and benefits (similar to the EU expert report). Writers should include discussion of any critical issues that arose during the development program that may have affected the outcomes and should be prepared to explain those issues based on supportive data elsewhere in the submission. The overview should only be about 30 pages long, depending of course on how complex the application is. To keep the document within the page limit, writers should refer the reviewers to the longer clinical summaries (Module 2.7) and Module 5 clinical study reports as needed for supporting details, and writers should tabulate conclusions and interpretations of data or information without repeating them, for more concise presentation. It is also helpful to discuss the issues by topic rather than study by study (eg, to describe pertinent nonclinical or quality issues that have potential clinical impact) [2, 8].

The targeted audience for the clinical overview is primarily the clinical reviewers; however, authors should be aware that those reviewing other sections of the dossier may use it as a reference to gain an understanding of the clinical implications. Regulatory writers should refer to the ICH M4E Guideline for greater detail on what should be included in the clinical overview. Table 4 provides suggestions for the clinical overview outline [8, 13].

Table 4. Potential topics for inclusion in the clinical overview [8, 13]

Clinical overview outline	Potential subheadings
1. Product development rationale	• Indications and usage • Dosage and administration • Dosage forms and strengths • Description
2. Overview of biopharmaceutics	• Dosage/form/strength proportionality • Food effects
3. Overview of clinical pharmacology	• Introduction • Pharmacokinetics • Pharmacodynamics and mechanism of action • Conclusions
4. Overview of efficacy	• Study designs • Efficacy endpoints • Study populations • Statistical methodology • Compliance • Dose selection • Efficacy results • Onset of action • Long-term benefits • Efficacy conclusions

(continued)

Table 4 (continued)

5. Overview of safety	• Safety plan • Nonclinical information related to safety • Extent of exposure • Treatment-emergent adverse events • Adverse drug reactions • Comparison of adverse event profile with other drugs in the class • Deaths and other serious adverse events • Laboratory evaluations, vital signs, and other safety evaluations • Potential effects on ECG and QTc interval • Safety in special populations • Drug interactions • Effects on ability to drive or operate machinery or impairment of mental ability • Worldwide marketing experience • Safety conclusions
6. Benefits and risks conclusions	• Contraindications • Warnings and precautions • Drug interactions • Patient counseling information
7. References	• References

The clinical summaries

Module 2.7 clinical summaries should be divided into four distinct, individual documents, and the guideline stipulates that together they should not exceed 400 pages. The exception to the page limit is for a product application that includes research on multiple indications; in this case, a separate summary of clinical efficacy should be written for each indication [8]. The clinical summaries should summarize all of the clinical research activities performed for product development, using the clinical study reports included in Module 5 as the source documents. As in the clinical overview, writers should refer liberally to the actual study reports in Module 5 to avoid lengthy data presentations that are available to the reviewer in full detail elsewhere. Including meaningful tables and graphs within the text that summarize the pertinent information concisely is encouraged by reviewers. Sponsors are discouraged from including any datasets in Module 2 with the clinical summaries in addition to those provided in Module 5 because such detail is usually associated with a detailed report rather than with a summary document.

Modules 2.7.1, 2.7.2, and 2.7.3 (summaries of biopharmaceutical, clinical pharmacology, and clinical efficacy studies) have similar recommended content, including a brief background and overview, a summary of results of all the applicable studies, and a comparison of the results across applicable studies. The summary of clinical safety (Module 2.7.4) focuses on drug exposure; adverse events; clinical laboratory evaluations; vital signs, physical findings, and other observations related to safety; safety in special groups and situations; and postmarketing data. Authors should avoid simple reiteration of the details of all of the studies by summarizing the findings and referring the reviewer to Module 5 for those details.

Module 5 contents
All the clinical study reports should be provided in Module 5 of the CTD, divided and placed in the appropriate section of Module 5 according to the type of study, as outlined in the ICH M4E Guideline (eg, biopharmaceutical, pharmacokinetic, or efficacy and safety studies). Chapter 6 (Clinical study reports) provides guidance to regulatory writers on how to approach writing a clinical study report.

Two documents should be placed in Module 5.3.5.3 that are unique to the US FDA requirements, according to 21 CFR 314.50(d)(5)(v) and 21 CFR 314.50(d)(5) (vi). The FDA requires pooling of clinical data, if possible, and inclusion of written reports that present the results of those integrated analyses; those reports are commonly referred to as the integrated summary of safety (ISS) and the integrated summary of efficacy (ISE) (Chapter 9, Integrated summaries of safety and efficacy). The FDA has clarified the requirements for these documents in industry presentations and by providing additional guidance [14]. Some confusion has resulted from the fact that the ISS and ISE are also referred to as summaries, but the FDA has made it clear that the intent of those documents is for sponsors to write reports detailing the sponsor's interpretation of the pooled data. The key results and conclusions that are detailed in the Module 5.3.5.3 reports of the integrated analyses (ISS and ISE) should be summed up in the Module 2 clinical summaries of safety and efficacy, similar to the way all the other clinical reports are summarized (ie, the Module 2 summaries are not interchangeable with the Module 5 reports). In fact, both are required per the US Federal Code of Regulations [11]. In some circumstances, a sponsor may be able to include enough information in the Module 2 summary to meet the regulation for including the ISS and ISE, but sponsors should be proactive in discussing the document strategy with the agency in advance to avoid being asked to supply additional documents after the dossier is submitted [2].

Conclusions

Just as the entire CTD can be portrayed in a pyramid fashion (with the primary source documents as the foundation that builds upward to the complete dossier), authors should approach writing the clinical documents with that pyramid hierarchy in mind. Documents in the CTD are related and dependent on each other and get smaller as they are built from the bottom up. Data and text from the foundation source documents must be finalized before the higher-level documents can be finalized, in this order:

1. Module 5 Clinical Study Reports
2. Module 2 Clinical Summaries
3. Module 2 Clinical Overview
4. Product Label

Each of these documents can be started before the predecessor is finalized, but additional quality control processes are imperative if draft data or text are entered into any of the higher-level documents before the lower-level documents are finalized. Spending a considerable amount of time on advance planning of writing these documents and including a skilled project manager on the submission team are the keys for successfully completing the clinical submission with quality results [2].

References

1   United States Government Accountability Office Report to Congressional Requesters. New Drug Development: Science, Business, Regulatory, and Intellectual Property Issues Cited as Hampering Drug Development Efforts. November 2006. Available at http://www.gao.gov/new.items/d0749.pdf (Accessed 20 February 2008).

2   Boe, P., Authoring the CTD: Leveraging Granularity, Reusability, and Simplicity. Image Solutions, Inc.; Whippany, NJ. Version 1.0; 2008.

3   Food and Drug Administration. Milestones in US Food and Drug Law History. Available at http://www.fda.gov/oc/history/default.htm (accessed 20 February 2008).

4   Zolman, T., An Introduction to eCTD: Electronic Common Technical Documents. Image Solutions, Inc.; Whippany, NJ. Version 1.0; October 2007.

5   International Conference on Harmonisation (ICH). M4 – Organisation of the Common Technical Document for the Registration of Pharmaceuticals for Human Use. Revision 3; January 2004. Available at http://www.ich.org/cache/compo/276-254-1.html (Accessed 21 February 2008).

6   International Conference on Harmonisation (ICH). The Common Technical Document for the Registration of Pharmaceuticals for Human Use: Quality – M4Q. The Quality Overall Summary of Module 2 and Module 3 Quality. Revision 1, September 2002. Available at http://www.ich.org/cache/compo/276-254-1.html (Accessed 21 February 2008).

7   International Conference on Harmonisation (ICH). The Common Technical Document for the Registration of Pharmaceuticals for Human Use: Safety – M4S. The Nonclinical Overview and Nonclinical Summaries of Module 2 and Organisation of Module 4. Revision 2, December 2002. Available at http://www.ich.org/cache/compo/276-254-1.html (Accessed 21 February 2008).

8   International Conference on Harmonisation (ICH). The Common Technical Document for the Registration of Pharmaceuticals for Human Use: Efficacy – M4E. Clinical Overview and Clinical Summary of Module 2 and Module 5 Clinical Study Reports. Revision 1, September 2002. Available at http://www.ich.org/cache/compo/276-254-1.html (Accessed 21 February 2008).

9   International Conference on Harmonisation (ICH). M2 EWG. Electronic Common Technical Document Specification V 3.2. February 04, 2004. Available at http://estri.ich.org/eCTD/eCTD_Specification_v3_2.pdf (Accessed 19 February 2008).

10 ICH Home Page: Structure of ICH. Available at http://www.ich.org/cache/compo/276-254-1.html (Accessed 20 February 2008).

11 The ICH Website. Available at http://www.ich.org/cache/compo/276-254-1.html (Accessed 21 February 2008).

12 FDA Centers for Drug and Biologics Evaluation and Research (CDER and CBER). Guidance for Industry: Providing Regulatory Submissions in Electronic Format – Human Pharmaceutical Product Applications and Related Submissions Using the eCTD Specifications. April 2006. Available at http://www.fda.gov/cder/guidance/7087rev.pdf (accessed 21 February 2008).

13 ISIWriter™ CTD Templates. Copyright Image Solutions, Inc. 2004.

14 FDA Centers for Drug and Biologics Evaluation and Research (CDER and CBER). Guidance for Industry: Integrated Summaries of Effectiveness and Safety: Location Within the Common Technical Document. Draft, June 2007. Available at http://www.fda.gov/cder/guidance/7621dft.pdf (accessed 21 February 2008).

Targeted Regulatory Writing Techniques. Clinical Documents for Drugs and Biologics,
edited by Linda Fossati Wood and MaryAnn Foote
© 2009 Birkhäuser Verlag Basel/Switzerland

Chapter 12.

Clinical trial procedures and approval processes in Japan

Takumi Ishida[1] and Katsunori Kurusu[2]

[1]Takumi Ishida, Japan Medical Linguistics Institute, Kobe, Japan;
[2]Katsunori Kurusu, Marketed Products Regulatory Affairs, Sanofi-Aventis KK, Tokyo, Japan

Laws and agencies for conducting clinical trial and approvals

Pharmaceutical affairs-related laws and regulations

Pharmaceutical administration in Japan is governed primarily by the Pharmaceutical Affairs Law (PAL; Law No. 145 issued in 1960) that controls clinical research, manufacturing, marketing, labeling, and safety of drugs, diagnostics, and medical devices (from here on referred to collectively as drugs). The need for a clinical trial notification (CTN) submission to the governing agency before human use of a new drug, for manufacturing/distribution approval application (MAA) before manufacturing/marketing of a new drug product, and for reliability/compliance review on application of a dossier is defined in Articles 80 (2) and 14, respectively, of the PAL. Procedures for implementing the PAL rules are the Enforcement Regulations of PAL (ER-PAL, Ministerial Ordinance No. 1 issued in 1961). Table 1 presents the ER-PAL articles and the laws they describe.

Table 1. PAL articles relevant to submissions

Article number	Title
268	Cases Requiring Notification of Clinical Trial on Medicinal Substances
269	Notification on Clinical Trial Plan on Medicinal Substances
38	Application for Marketing Approval of Drugs, etc
40	Data to be Attached to Applications for Approval
43	Standards for Reliability of Application Data

Based on these laws, regulations related to pharmaceutical administration are issued by the Ministry of Health, Labour and Welfare (MHLW) in the form of a ministerial ordinance and by the Pharmaceutical and Food Safety Bureau (PFSB) in the form of notifications. Pharmaceutical laws, ministerial ordinances, and notifica-tions are issued by bureaus of the MHLW such as PFSB. Laws are required to be approved by the Diet but other issues do not have this requirement.

Governing agency: MHLW and PFSB

The agency that has authority to approve drugs is the MHLW. Daily practices (ap-proval and license authority) for pharmaceutical affairs such as clinical trials, ap-proval review, and postmarketing surveillance are handled by one of its 11 divisions, PFSB. The divisions of the PFSB governing pharmaceutical administration are the Evaluation and Licensing Division, Safety Division, Compliance and Narcotics Divi-sion, and Blood and Blood Products Division [1].

Governing agency: PMDA

Review of a CTN and an MAA is undertaken not by the PFSB but by the inde-pendent administrative institution, Pharmaceuticals and Medical Devices Agency (PMDA; the Agency in this chapter); therefore, the applications are submitted not to the Evaluation and Licensing Division of PFSB but the PMDA. The PMDA is not concerned with the issue of laws, ministerial ordinances, or notifications but under-takes the practical implementation of these regulations [2].

Preparation for clinical trials

Preparation for conducting a clinical trial starts with the sponsor requesting a clini-cal trial consultation under the interview advice or face-to-face advice system. The sponsor may request a prior consultation with the PMDA to organize consultation items so that the clinical trial consultation proceeds smoothly.

Consultation system
The consultation system in Japan allows sponsors to discuss with the PMDA matters related to clinical trials and approval applications (Table 2).

Table 2. Consultations related to new drug development

Consultation	Example
Consultation on application procedures	Procedures necessary for initiating clinical trials and data required for CTN
Consultations on bioequivalence studies	Issues relating to the use of foreign data when the formulation to be marketed in Japan is different from the formulation for foreign markets; decision if an application is to be classified as a new formulation application or generic drug application; validity of parameters to be used in bioequivalence studies; rationale for concluding bioequivalence between formulations based on results of bioequivalence studies
Consultations on drug safety	Evaluation of animal data that raise concerns about carcinogenic potential; safety evaluation of new excipients
Consultations on drug quality	Specifications/test methods for biotechnology-derived pharmaceuticals; specifications/test methods for controlled-release dosage forms and kit products
Consultations before start for phase 1 studies	The type and extent of nonclinical data necessary for the conduct of human trials; identification of an initial starting dose and a dose-increment regimen in phase 1 studies (including studies of anticancer drugs); acceptability of foreign phase 1 studies in support of CTN; adequacy of information contained in informed consent
Consultations before start of early phase 2 studies	Validity of pharmacokinetic parameters to be used in early phase 2 studies
Consultations before start of late phase 2 studies	Dose levels in phase 2 studies; adequacy of information contained in informed consent
Consultations after completion of phase 2 studies	Rationale for interpreting dose-response data and selecting recommended clinical dose; selection of a comparator(s) and endpoints and statistical analysis plan in phase 3 studies; requirement of a study or studies other than controlled studies; adequacy of information contained in informed consent
Pre-application consultations	Preparation of clinical study reports and application summary (CTD Module 2); identification of clinical data supportive of approval application
Consultations when planning clinical trials for re-examinations and re-evaluations	Guidance and advice on the plan of postmarketing clinical studies for re-examination and re-evaluation applications
Consultations on completion of clinical trials for re-examinations and re-evaluations	Guidance and advice on the preparation of application package and adequacy of data to be contained in the package after or close to the completion of clinical studies for re-examination and re-evaluation applications. Preparation and formatting of clinical study reports; identification of clinical data supportive of re-examination and re-evaluation applications

(continued)

Table 2 (continued)

Consultation	Example
Additional consultations	Consultations after the initial consultation before start of phase 1 studies until the consultation before start of phase 2 studies; consultations after the consultation before start of phase 2 studies until the consultation after completion of phase 2 studies; consultations after the consultation after completion of phase 2 studies until the pre-application consultation; consultations after the pre-application consultation until filing; consultations after the consultation for planning clinical trials for re-examination and re-evaluation until the consultation at completion of clinical trials for re-examination and re-evaluation; consultations after the consultation at completion of clinical trials for re-examination and re-evaluation until the completion of re-examination and re-evaluation by the Ministry; consultations on exclusively quality-related issues after the initial consultation on quality; consultations on exclusively safety-related issues after the initial consultation on safety

Data necessary for a CTN

The requirements for a CTN are specified in the Standards for the Conduct of Clinical Trials of Pharmaceutical Products (Good Clinical Practice Guideline: Ordinance No. 28 dated March 27, 1997). Procedures for the application are stated in the Enforcement Regulations of the Standards for the Conduct of Clinical Trials of Pharmaceutical Products [Communication No. 430 of the Pharmaceutical Affairs Bureau (PAB, currently PMSB) dated March 27, 1997]. The CTN does not need to be submitted in the CTD format. Data to be attached to the initial CTN are listed in Table 3, and CTN Submission Form 7 is provided in Appendix XIV.

Table 3. Data to be included with the initial CTN

• Data showing that the clinical trial is scientifically appropriate
• Protocol for the planned study
• Patient information and consent form
• Sample case report form
• Latest version of the investigator's brochure

Points to consider in preparing CTN documents

The 'Document Stating the Reason that the Request for the Clinical Trial is Judged to be Scientifically Appropriate' to be attached to the CTN should be prepared based on the latest version of investigator's brochure, together with other data supporting the use of the investigational drug in humans. The document should be a concise summary (approximately three to five pages); however, its contents should be sufficiently detailed to allow the Agency to approve the trial. In particular, the sponsor must carefully examine and evaluate data and information to be included in the initial CTN.

Clinical protocols

In general, the format and contents of clinical protocols are the same for all three regions, the European Union (EU), Japan, and the United States. In developing the protocol, the sponsor must be careful to evaluate the potential influence of ethnic factors such as physical constitution, domestic clinical guidelines, and differences in standards/normal values of measurements that may affect the study outcome. The Agency may have questions concerning the protocol; typical examples are provided in Table 4.

Table 4. Typical questions asked by agency

Introduction/background:
• Why the clinical trial is scientifically appropriate should be sufficiently explained.
Collaboration among persons or parties concerned:
• Requests and directions to investigators as well as information provided should be concretely described.
• If study conduct or any procedure is contracted to a contract research organization (CRO), the function and responsibility of the CRO should be clearly defined.
Rationale for dosage and administration:
• The reasons for the initial and highest doses selected should be explained by adequately referring to toxic and nontoxic doses in nonclinical studies, even if the project is proceeding in foreign countries. It is not appropriate to simply mention that "foreign clinical studies are conducted at such-and-such doses."
• The rationale for selected dosage should be explained with pharmacokinetic evidence showing no difference between foreign and Japanese populations, even if the dosage is the same between regions.
• The rationale for the dosage should be based on foreign safety information.
• The rationale for moving into the later stages and the appropriateness of the timing of safety evaluations should be explained.
Rationale for outcomes and time of observations/examinations:
• The reasons for selecting outcomes and time of observations/examinations should be clearly explained. For example, the time of observation/examination should be consistent with $t_{1/2}$ data.
• When the protocol is designed based on foreign data, racial differences in pharmacokinetics tend to pose problems. It is advised that the metabolic fate, type of P450 isoforms involved, etc, be confirmed to be identical between races.
Criteria for subject inclusion/exclusion, dose increase, and treatment termination:
• Inclusion/exclusion criteria should be established to ensure subject safety by evaluating potential risks anticipated based on results from available nonclinical and clinical data.
• "Subjects judged to be healthy by the physician" should be clearly explained.
• Data to be used for dose escalation, time schedule of assessment, assessment criteria, and discontinuation criteria should be clear. For example, discontinuation criteria should be made clear by stating "the treatment is discontinued when such-and-such adverse events are encountered."
Statistical analysis:
• Statistical analysis should be scientifically adequate and suitable for the purpose of the study.

(continued)

Table 4 (continued)

Safety:
- If certain adverse events are not defined as the reasons for discontinuation, the criteria for continuation or discontinuation and reason for continuation should be described.
- The incidence and causality relationship of adverse events in foreign studies should be described in detail.
- When a bridging study is planned according to the ICH E5 guideline, by referring to foreign clinical data, the protocol design should show that ethnic factors are taken into account for ensuring subject safety.
- When a genetically engineered drug is used in the trial, the safety of the study drug should be confirmed by attaching quality data (eg, purity/impurity, contamination) prepared based on "Specifications: Test Procedures and Acceptance Criteria for Biotechnological/Biological Products."

Informed consent:
- Laboratory tests that require consent of subjects should be stated in informed consent. For example, "blood is sampled in a quantity of ×××mL for such-and-such tests."
- Risks (safety) for subjects should be adequately described.
- Relevant guidelines state that the sponsor is responsible for requesting that the investigator prepare an informed consent form.

Coordinating investigator

The name of the coordinating (principal) investigator must be provided on the CTN form. The coordinating investigator in a multicenter study is the physician who is responsible for ensuring that all sites follow the protocol and other details of the trial. The coordinating investigator may be the study coordinating committee.

Guidelines for clinical evaluation of drugs

As of March 2008, 29 guidelines for the clinical evaluation of drugs have been published; 18 of these are Japanese, not ICH, guidelines (Table 5). These guidelines may be referred to in developing a study protocol.

Table 5. Guidelines for the clinical evaluation of drugs

- Guidelines on Clinical Evaluation Methods of Oral Contraceptives (Notification No. 10 of the First Evaluation and Registration Division, PAB dated April 21, 1987).
- Guidelines on Clinical Evaluation Methods of Drugs to Improve Cerebral Circulation and/or Metabolism in Cerebrovascular Disorders (Notification No. 22 of the First Evaluation and Registration Division, PAB dated October 31, 1987).
- Guidelines on Clinical Evaluation Methods of Antihyperlipidemic Drugs (Notification No. 1 of the First Evaluation and Registration Division, PAB dated January 5, 1988)
- Guidelines on Clinical Evaluation Methods of Antianxiety Drugs (Notification No. 7 of the First Evaluation and Registration Division, PAB dated March 16, 1988).
- Guidelines on Clinical Evaluation Methods of Hypnotics (Notification No. 18 of the First Evaluation and Registration Division, PAB dated July 18, 1988).

(continued)

Table 5 (continued)

- Guidelines for Clinical Evaluation Methods of Antibacterial Drugs (Notification No. 743 of the New Drugs Division, PMSB dated August 25, 1998).
- Guidelines on Clinical Evaluation Methods of Drugs to Treat Heart Failure (Notification No. 84 of the First Evaluation and Registration Division, PAB dated October 19, 1988).
- Guidelines on Clinical Evaluation Methods of Drugs to Treat Osteoporosis (Notification No. 742 of the Evaluation and Licensing Division, PMSB dated April 15, 1999)
- Guidelines on Clinical Evaluation Methods of Antiarrhythmic Drugs (Notification No. 0325035 of the Evaluation and Licensing Division, PFSB dated March 25, 2004)
- Guidelines on Clinical Evaluation Methods of Antianginal Drugs (Notification No. 0512001 of the Evaluation and Licensing Division, PFSB dated May 12, 2004)
- Guidelines for Clinical Evaluation Methods of Antimalignant Tumor Drugs (Notification No. 1101001 of the Evaluation and Licensing Division, PFSB dated November 1, 2005).
- Guidelines for Clinical Evaluation Methods of Antirheumatoid Drugs (Notification No. 0217001 of the Evaluation and Licensing Division, PFSB dated February 17, 2006).
- Guidelines for Clinical Evaluation Methods of Drugs for Overactive Bladder or Incontinence (Notification No. 0628001 of the Evaluation and Licensing Division, PFSB dated June 28, 2006).
- Research on Evaluation Methods of Immunotherapeutic Agents for Malignant Tumors (1980).
- Research on Evaluation Methods of Blood Preparations, Especially Plasma Fraction Preparations (1984).
- Research on Overall Evaluation Methods of Interferon Preparations (1984).
- Guidelines on Clinical Evaluation Methods of Anti-inflammatory Analgesic Drugs (1985).
- Guidelines on the Design and Evaluation of Sustained-release (Oral) Preparations (Notification No. 5 of the First Evaluation and Registration Division, PAB dated March 21, 1988).

Submission of a CTN and start of clinical trials

A CTN should be submitted to the PMDA on a computer disc. A contract with the investigational sites participating in the first trial must be made at least 30 days after the Agency has received the CTN, which allows 30 days for interactions between the sponsor and the Agency. A number of items must be submitted to investigational sites:

- Protocol
- Sample case report form
- Informed consent
- Document explaining compensation to the subjects in the event of trial-related injuries
- Document pertaining to clinical trial cost estimate (including a document pertaining to payment to study subjects)
- Contract agreement (if available)
- Summary of the protocol and study drug
- Standard operating procedures for drug control
- Check list (for use at a hearing to confirm study design or procedures with the sponsor)

When the trial is completed, the clinical study report is prepared according to the ICH E3 guideline [3].

Notifications to be submitted during clinical development

CTN for trial start, protocol modification, and trial termination

The CTN for the second and all subsequent studies for a given drug is submitted to the PMDA approximately 2 weeks before the estimated date of each contract with the investigative site using Form 9 (Appendix XIV) and supportive data (Table 6).

Table 6. Information to be submitted for subsequent trials

- Data showing that the clinical trial is scientifically appropriate (including study results and other new information obtained after the previous CTN)
- Protocol
- Written patient information and consent form (one copy of documents if different documents are used at investigational sites in a trial)
- Sample case report form
- Latest version of the investigator's brochure

If the protocol is modified while the trial is ongoing, the PMDA must be notified of every modifications through a CTN either before or after implementation. If changes are made to the study objectives or the target disease, then notification with a new CTN (instead of a change notification) is required.

The PMDA must be notified of changes approximately 6 months before implementation of the changes (Table 7). As a rule, other changes require notification before implementation of the changes. Notification of trial discontinuation and termination must be submitted to the Agency using Forms 11 and 13, respectively (Appendix XIV).

Table 7. Changes that must be made known to PMDA

- Deletion of the name of medical institution(s), coordinating investigator, and member(s) of the study coordinating committee
- Changes regarding subinvestigator(s) (eg, addition, deletion, title) and change in the title of investigator(s), coordinating investigator, and member(s) of the study coordinating committee
- Changes in the amount of clinical sample to be delivered to investigational sites and planned size of study population
- Slight change in study period due to differences in contract dates among institutions. (Notification is also necessary when the latest date of last patient observation is delayed.)
- Changes in the name of medical institution, participating hospital department, and address/phone number of institution
- Change in person responsible for trial conduct/management at institution

---

Reporting of adverse events

Any adverse reactions to an investigational drug and any infections occurring during the clinical trial are required to be promptly reported to the Minister of MHLW according to the procedures of reporting specified in Article 273 of the ER-PAL. Reporting of adverse reactions and infections is mandatory after termination of the trial and throughout the review period of the approval application. The reporting system during clinical trials is basically harmonized with ICH guidelines. The reporting procedures are defined in regulations Clinical Safety Data Management (Notification No. 227 of the Pharmaceuticals and Cosmetics Division, PAB dated March 20, 1995); Reporting of Adverse Reactions, Etc to the PMDA (Notification No. 0330001 of PFSB dated March 30, 2004); Points to Consider for Approval Application Data for New Drugs (Notification Nos. 0331022 of the Evaluation and Licensing Division and 0331009 of the Safety Division, PFSB dated March 31, 2006); Points to Consider in Reporting Adverse Reactions (Notification No. 0426001 of the Evaluation and Licensing Division, PFSB dated April 26, 2006); and Q and A on Adverse Reaction Reporting (Office Communication dated May 31, 2006).

The electronic reporting procedures based on ICH E2B are detailed in Data Elements and Interchange Formats for Transmission of Individual Case Safety Reports (Notification Nos. 344 of the Evaluation and Licensing Division and 39 of the Safety Division, PFSB dated May 31, 200) and Q and A on Adverse Reaction Reporting (Office Communication dated May 31, 2006). The scope of safety information to be reported in cases of adverse reactions is provided in Table 8.

Table 8. Scope of safety information for adverse reactions

---

1. Cases of death or potentially leading to death associated with causes unpredicted from the investigator's brochure – 7-day reports (Article 273-(1) of the ER-PAL). These are cases that are thought to result from adverse reactions of, or are attributable to, infectious diseases suspected to be associated with the use of the investigational drug or a drug used in a foreign country that is confirmed to be equivalent in ingredients with the investigational drug. The frequency of the occurrence, including the number of occurrences, the rate, and the conditions of occurrence, cannot be predicted from the investigator's brochure.
2. The following cases (excluding those specified in the preceding item) – 15-day reports (Article 273-(2)-(A) of the ER-PAL).
   2.1. The following cases that are suspected to result from adverse reactions of, or are attributable to, infectious diseases suspected to be associated with the use of the investigational drug of which the tendency of the occurrence, including the number of occurrence, the incidence, and the conditions of occurrence, cannot be predicted from the investigator's brochure
      2.1.1 Cases requiring hospitalization or prolongation of hospital stay for treatment
      2.1.2 Disabling
      2.1.3 Cases potentially leading to disabling
      2.1.4 Severe cases comparable with those specified in (2.1.1) to (2.1.3) inclusive and with death or potentially leading to death of the preceding item
      2.1.5 Congenital disease or abnormality in the subsequent generations
   2.2 The occurrence of cases of death or potentially leading to death (Article 273-(2)-(B) of the ER-PAL). Cases of death or potentially leading to death of the preceding item , which are suspected to be due to adverse reactions of, or are attributable to, infectious diseases suspected to be associated with the use of the investigational drug
   2.3. Safety measures taken in foreign countries (Article 273-(2)-(C) of the ER-PAL). Safety measures taken to prevent the occurrence or spread of hazards to the public health such as discontinuation of manufacture, import or retail, or recall or disposal of a drug used in a foreign country that is confirmed to be equivalent in ingredients with the investigational drug
3. Research reports [Article 273-(2)-(D) of the ER-PAL]. Research reports showing the potential risk to cause cancer or other serious disease, disabling or death due to adverse reactions of, or infectious diseases associated with the use of, the investigational drug, which shows that substantial changes have been observed in the tendency of the occurrence, including the number of cases, incidence, conditions of occurrence, etc or those showing the lack of the anticipated efficacy or clinical benefit of the investigational drug for the disease subject to clinical trials

---

International collaborative studies

In view of the current trend of increasing the number of international studies, the MHLW issued a guidance that includes the Agency's position on the issues of ethnic factors and study design [4].

Preparation for manufacturing/marketing approval application

It is recommended that the sponsor request a pre-application consultation with the Agency after or close to the completion of clinical research to obtain guidance and advice on the preparation of the application package, the adequacy of the data to be contained in the package, and the acceptability of the interpretation of the clinical data.

Marketing approval application dossier

Article 14-(1) of the PAL requires a sponsor that intends to market a drug to obtain marketing approval of the Minister, and Article 14-(3) of the PAL requires the sponsor to include data related to the results of clinical trials and other pertinent data as specified by the ER-PAL.

Application form
Application Form 22 (1), used for manufacturing/marketing approval applications, is provided in Appendix XIV.

Data to be included in application for approval
Data to be submitted to the Agency for approval review are specified in Article 40 of the ER-PAL and are provided in Table 9. Section A "Origin and Background of the Discovery as Well as the Conditions of Use in Foreign Countries" must contain information on the properties and characteristics of the drug and comparison with other drugs in class. This information is used not only for evaluation but also for pricing. Application data are specified in Notification No. 0331015 of the PMDA entitled "Data Required for Applications for Prescription Drugs" dated March 31, 2005. Table 10 summarizes the MAA data required.

Table 9. Data to be submitted to the agency for approval review

---

- Origin and background of the discovery as well as the conditions of use in foreign countries
- Manufacturing methods, specifications, and test methods
- Stability
- Pharmacology
- Absorption, distribution, metabolism, and excretion
- Acute/subacute/chronic toxicity, teratogenicity, and other toxicity
- Clinical studies

Table 10. Data to be submitted for approval to manufacture/distribute a new prescription drug

	Data and drugs	New drug	New combination	New route	New indication	New formulation	New dosage	Additional formulation	Similar composition	Other
a	Origin	●	●	●	●	●	●	●	●	×
	Foreign country	●	●	●	●	●	●	●	●	×
	Comparison	●	●	●	●	●	●	●	●	×
b	Structure	●	×	×	×	×	×	×	×	×
	Manufacture	●	●	●	×	●	×	●	●	△
	Specifications	●	●	●	×	●	×	●	●	●
c	Long-term	●	●	●	×	●	×	△	●	×
	Severe	●	●	●	×	●	×	△	●	×
	Accelerated	●	●	●	×	●	×	●	●	●
d	Efficacy	●	●	●	●	×	●	×	△	×
	Safety	●	△	△	×	×	×	×	△	×
	Other	△	△	△	×	×	×	×	×	×
e	Absorption	●	●	●	△	●	●	×	×	×
	Distribution	●	●	●	△	●	●	×	×	×
	Metabolism	●	●	●	△	●	●	×	×	×
	Excretion	●	●	●	△	●	●	×	×	×
	Bioequivalence	×	×	×	×	×	×	●	×	●
	Other	△	△	△	△	△	△	△	×	×

f	**Single dose**	•	•	•	×	×	×	×	×
	Repeated dose	•	•	•	×	×	×	∆	×
	Genotoxicity	•	×	×	×	×	×	×	×
	Carcinogenicity	∆	×	∆	×	×	×	×	×
	Reproductive toxicity	•	×	•	×	×	×	×	×
	Local irritation	∆	∆	∆	×	×	×	∆	×
	Other	∆	×	∆	×	×	×	×	×
g	**Clinical**	•	•	•	•	•	×	•	×

•: Date required; ×: Data not required; ∆: Data required depending on individual cases

Format of application

The CTD format (Notification No. 899 of PMSB dated June 21, 2001 and No. 0701004 of PFSB dated July 1, 2003, and Office communication dated October 22, 2003) is used for submission to the Agency for approval of a new drug. The region-specific Module 1 (regulatory information and draft package insert) of the CTD dossier for Japan contains the information provided in Table 11. The product overview document for an application was previously called a 'Gaiyo,' but is now prepared according to Module 2 of the ICH M4 guideline (Table 12).

Table 11. Region-specific Module 1 of the CTD dossier for Japan

- Table of contents
- Approval application (copy)
- Certificates (declarations of those responsible for collection and compilation of data for approval applications, GLP- and GCP-related data, contracts for co-development, etc)
- Patent status
- Background of origin, discovery, and development
- Data related to conditions of use in foreign countries
- List of related products
- Package insert (draft)
- Documents concerning nonproprietary name
- Data for review of designation as poisons, deleterious substances, etc
- Draft of basic protocol for postmarketing surveillance
- List of attached documentation
- Other

Table 12. Mapping the Japanese marketing approval application to the CTD modules

Data		CTD module
A. Origin and background of the discovery as well as the conditions of use in foreign countries	1. Origin, discovery and development	1
	2. Conditions of use in foreign countries, etc	1
	3. Data on related products	1
B. Manufacturing methods, specifications, and test methods	1. Structure	2.3 and 3
	2. Manufacture	2.3 and 3
	3. Specifications and test methods	2.3 and 3
C. Stability	1. Long-term storage conditions	2.3 and 3
	2. Severe storage conditions	2.3 and 3
	3. Accelerated storage conditions	2.3 and 3
D. Pharmacology	1. Efficacy pharmacology	2.4, 2.6, 4
	2. Safety pharmacology	2.4, 2.6, 4
	3. Other pharmacology	2.4, 2.6, 4
E. Absorption, distribution, metabolism, and excretion	1. Absorption	2.4, 2.6, 4
	2. Distribution	2.4, 2.6, 4
	3. Metabolism	2.4, 2.6, 4
	4. Excretion	2.4, 2.6, 4
	5. Bioequivalence	2.4, 2.6, 4
	6. Other	2.4, 2.6, 4
F. Acute/subactute/chronic toxicity, teratogenicity, and other toxicity	1. Single-dose toxicity	2.4, 2.6, 4
	2. Repeated-dose toxicity	2.4, 2.6, 4
	3. Genotoxicity	2.4, 2.6, 4
	4. Carcinogenicity	2.4, 2.6, 4
	5. Reproductive toxicity	2.4, 2.6, 4
	6. Local irritation	2.4, 2.6, 4
	7. Other	2.4, 2.6, 4
G. Clinical studies	Clinical studies	2.5, 2.7, 5

Submission of application

The application form and data are submitted to the PMDA. The flow of the approval process is shown in Figures 1 and 2.

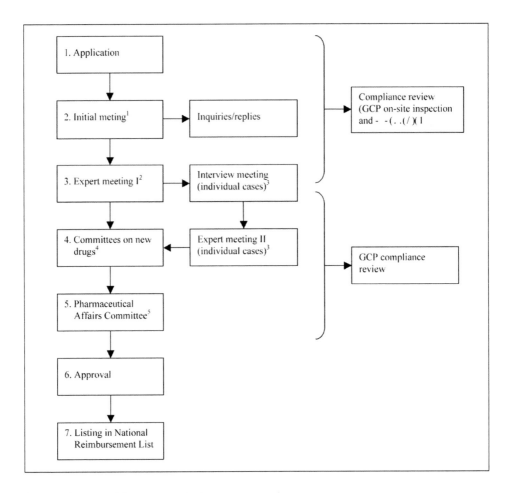

Figure 1. Approval of Japanese application.

[1]*Initial meeting:*

PMDA organizes review teams for individual applications. At the initial meeting, the team receives briefing on the drug from the applicant, such as characteristics of the drug, history of development, and structure of the application dossier. The applicant may answer inquiries raised earlier by the team, argue about issues the applicant has concerns about, or receive new inquiries. The team prepares a review report (Review Report I: Summary of application data and outline of review) after the meeting.

[2]*Expert Meeting I:*

The review team requests experts to review the application dossier together with the Review Report I and discuss important problems with the experts for the coordination of opinions. An Expert Meeting II may not be held until major problems are solved.

[3]*Interview Meeting:*

The Review Report is sent to the applicant before the Interview Meeting. The meeting attended by the applicant, its experts, review team, and its experts discuss major pending issues and problems. If necessary, an Expert Meeting II may be held. Results of discussions are summarized as Review Report II.When the issues and problems are solved, the final review report is prepared, attaching Review Reports I and II.

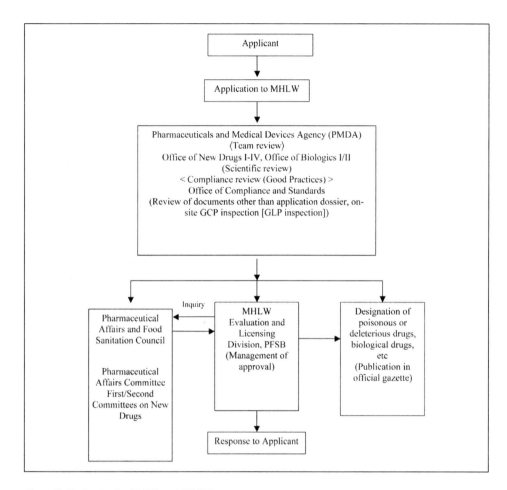

Figure 2. Review by the PMDA and MHLW.

Figure 1 (continued)
[4]*Committees on New Drugs:*
The First Committee on New Drugs (antibiotics, chemotherapies, antimalignant tumor drugs, blood products, and biologic products) and the Second Committee on New Drugs (other drugs) review applications by drug category based on the consultative document from the PMDA, review reports from the agency, application forms, a draft package insert, Module 2 documents, a list of attached documents, and Expert Meeting reports. Members of these committees are not staff of the PMDA.
[5]*Pharmaceutical Affairs Committee:*
Experts in medical and pharmaceutical sciences examine and review important pharmaceutical matters based on review report from the Committees on New Drugs, review reports from the Agency, draft package insert, and a list of participating medical institutions.

Drug master file system

The goal of the drug master file system is both to protect intellectual property and to facilitate review by allowing a registrant (master file registrant) other than an applicant to separately submit information on quality and the manufacturing method at the time of approval reviews of drug substances, new excipients, containers, and packaging materials to be used in drug products (revised PAL issued on April 1, 2005). Overseas manufacturers can use the drug master file system by appointing a drug substance manager.

Priority review

Drug approval reviews are normally processed in the order that the application forms are received, but drugs designated as orphan drugs and other drugs considered to be particularly important from a medical standpoint, can request a priority review designation based on an overall evaluation of the seriousness of the targeted disease and the clinical usefulness of the drug (Notification No. 0227016 of the Evaluation and Licensing Division, PFSB dated February 27, 2004).

Approval

Approval refers to governmental permission to market a drug that has demonstrated quality, efficacy, and safety or to market a drug that is manufactured by a method in compliance with manufacturing control and quality control standards based on an appropriate quality and safety management system. Approval allows the drug to be marketed, generally distributed, and used for healthcare in Japan.

Sponsors that wish to start a marketing business for drugs are required to obtain a marketing business license (manufacturing/distribution approval). The licensing requirements include the appointment of a general marketing compliance officer (eg, pharmacist) and compliance with the Good Quality Practice (GQP) and Good Vigilance Practice (GVP). The general marketing compliance officer, the quality assurance supervisor of the quality assurance unit in charge of GQP, and the safety management supervisor of the general safety management division in charge of GVP are known as the 'manufacturing/marketing triumvirate' and are at the center of the marketing system.

Postmarketing surveillance

The sponsor must submit a Draft of Basic Protocol for Postmarketing Surveillance at the time of the submission to agree that all processes will be taken to ensure the efficacy and safety of drugs after they are marketed. The postmarketing surveillance consists of three systems: the adverse drug reaction reporting system, the re-examination system, and the re-evaluation system.

ADR reporting system

Good Postmarketing Surveillance Practice (GPSP) requires compliance by manufacturers/distributors when performing postmarketing surveillance or studies, and also provides compliance criteria for the preparation of data for re-examination and re-evaluation applications (Table 13).

Table 13. Good postmarketing surveillance practice

- Early postmarketing surveillance: Safety assurance activities that are performed within 6 months after commencement of marketing to promote proper use of the drug in medical treatment and to quickly identify the occurrence of serious adverse reactions, etc It is specified as a condition of approval.
- Postmarketing surveys: Drug use-results surveys and postmarketing clinical studies that the distributor of drugs conducts to collect, screen, confirm, or verify information relating to the quality, efficacy, and safety of drugs.
  - Drug use-results surveys: Surveys to screen or confirm information related to the incidence of adverse reactions, together with the quality, efficacy, and safety of drugs, without specifying the condition of the patients that use the drugs
  - Specified drug-use surveys: Surveys to screen or confirm information relating to the incidence of adverse reactions together with the quality, efficacy, and safety of drugs in specified populations of patients such as pediatric patients, elderly patients, pregnant women, patients with renal and/or hepatic disorders, and patients using the drug for long periods
  - Postmarketing clinical studies: Clinical studies conducted in accordance with approved dosage and administration, and indications to collect information on quality, efficacy, and safety unobtainable in routine medical practice.
- Periodic safety update reports (PSUR): Periodic safety report system adopted on the basis of agreements at the ICH
- Other: The adverse reaction reporting system undertaken by pharmaceutical companies based on spontaneous reports of adverse reactions/infections received, the drug safety information reporting system undertaken by medical personnel, and the WHO International Drug Monitoring Program whereby drug safety information is exchanged among various countries.

Re-examination system

The re-examination system is aimed at reconfirmation of the clinical usefulness of drugs by collecting information on efficacy and safety under various clinical settings (eg, larger population, concomitant medication, complications, and age) during a specified duration after approval. The period for collecting such data (re-examination period) is specified for individual drugs from 4 to 10 years (eg, 8 years for drugs containing new active ingredients, 10 years for orphan drugs). The re-examination application package should include a summary of drug use-results surveys; specific drug-use survey reports; postmarketing clinical trial reports; data from patients who have developed adverse reactions or infections; data from research reports; reports of specific measures adopted in Japan and overseas; and reports of serious adverse reactions. The outcome of the review can be approval refused (ie, manufacturing and marketing suspended, approval revoked); changes in approval required; or approval.

Re-evaluation system

The re-evaluation system is a system whereby the efficacy and safety of not only proprietary but also nonproprietary drugs are re-evaluated on the basis of the current status of medical and pharmaceutical sciences.

References

1   Organization of the MHLW; http://www.mhlw.go.jp/english/org/detail/index.html (Accessed 11 March 2008).
2   Organization of the PMDA; http://www.pmda.go.jp/english/about/organization.html (Accessed 11 March 2008).
3   ICH E3, Guideline for Industry, Structure and Content of Clinical Study Reports, July 1996.
4   Notification No. 0928010 of the Evaluation and Licensing Division, PMSB dated September 28, 2007. http://www.pmda.go.jp/topics/h200110kohyo.html (Accessed 11 March 2008).

Targeted Regulatory Writing Techniques. Clinical Documents for Drugs and Biologics,
edited by Linda Fossati Wood and MaryAnn Foote
© 2009 Birkhäuser Verlag Basel/Switzerland

Chapter 13.

Region-specific submissions: United States of America

Linda Fossati Wood

MedWrite, Inc., Westford, Massachusetts, USA

Investigational new drug application

The purpose of an Investigational New Drug Application (IND) is to demonstrate that a product is reasonably safe for first-time use in humans. Several types of INDs exist, based on who is submitting (Investigator IND versus an IND submitted by a sponsor); the urgency of use of the investigational drug (Emergency use IND or Treatment IND); or the regulatory strategy (traditional IND versus Exploratory IND). This chapter describes a traditional IND, the most commonly used type and the IND most regulatory writers will prepare.

Evidence of safety is provided by submission in the IND of nonclinical test results, manufacturing information, testing performed on humans outside the United States (if applicable), and postmarketing information (also if applicable). The Food and Drug Administration (FDA) has 30 calendar days after receipt of the IND to review the information and decide whether the sponsor may proceed with clinical trials. If the FDA perceives either that the sponsor has not provided enough information to make this determination or that a drug that may pose an unreasonable risk to humans, then the FDA will issue a deficiency letter and put the sponsor's clinical trial program on 'hold.' Clinical hold means that the clinical trial may not enroll any subjects [1] (Figure 1). A study is placed on hold most frequently because of inadequate demonstration of safety based on nonclinical study results. The number of studies, the types of studies, the model used, or the results in general may not be considered by FDA reviewers to adequately predict safety. Other reasons for clinical hold are a study protocol that does not give FDA confidence that subjects will be protected from harm to the greatest degree possible or manufacturing methods that do not satisfy minimum criteria for drug quality and purity.

Clinical hold has serious consequences for drug development because the sponsor must address issues in the deficiency letter to the satisfaction of FDA before the

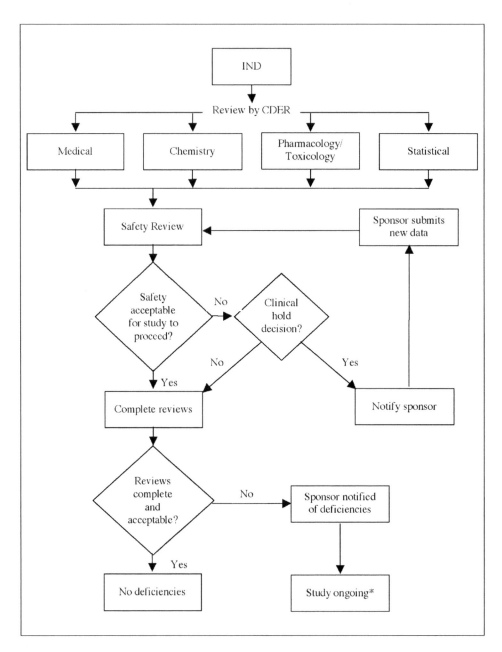

Figure 1. IND application process [1]
* while sponsor addresses deficiencies

clinical trial can start [2]. The period of time for this delay is indefinite, since it is based on negotiations between the sponsor and the FDA, and tends to cause great anxiety for the sponsor.

Good writing will never be able to overcome the perils inherent in an unsafe product (nor should writing try to overcome this). However, every product deserves a chance, since even drugs that initially seem of great risk and little benefit may show themselves to be powerful therapeutic tools, particularly when treatment options are limited or nonexistent. Therefore, the onus is on the writer to clearly convey, in a brief, succinct, and focused manner all available information to allow FDA to judge risk to humans. The writer must always be cognizant of the 30-day review period for the FDA.

The clinical sections of an IND (sections of an IND are called items) are presented in Table 1 as defined by the United States Code of Federal Regulations (CFR) [3]. Before introduction of the Commn Technical Document (CTD) format for IND submissions, INDs were written using the structure defined on Form FDA 1571 and described in Title 21 CFR Part 312. This outline is presented in Table 1.

Table 1. Outline of an IND as described on Form FDA 1571 and in Title 21 CFR Part 312

1. Form FDA 1571
2. Table of contents
3. Introductory statement
4. General investigational plan
5. Investigator's brochure
6. Protocol(s)
a. Study protocol
b. Investigator data
c. Facilities data
d. Institutional review board data
7. Chemistry, manufacturing, and controls data
Environmental assessment or claim for exclusion
8. Pharmacology and toxicology data
9. Previous human experience
10. Additional information

As part of the effort to harmonize global submission, and consistent with the idea that submissions should be built in such a way as to allow easy updates with new information, the structure of an IND is changing. IND content as defined in the items on the Form FDA 1571 will now be placed in the backbone of the CTD. Placement of IND item content into the appropriate sections of the CTD outline is presented in Table 2.

Table 2. Mapping clinical sections of the IND to CTD format [8]

CFR		CTD outline		
Number	Title	Module	Number	Title
312.23(a)(3)(i)	Item 3: Introductory statement	2	2.2	Introduction to summary
312.23(a)(3)(ii-iii)	Item 3: Introductory statement	2	2.5	Clinical overall summary
312.23(a)(3)(iv)	Item 4: General investigational plan	1	1.13.9	General investigational plan
312.23(a)(5)	Item 5: Investigator's brochure	1	1.14.4.1	Investigator's brochure
312.23(a)(6)	Item 6: Clinical protocol	5	5.3	Clinical protocol
312.23(a)(9)	Item 9: Previous human experience	2	2.5	Clinical overview
312.23(a)(9)	Item 9: Previous human experience	2	2.7	Clinical summary
312.23(a)(9)	Item 9: Previous human experience	5	5.3	Clinical study reports

The initial challenge of fitting information for an IND (a proposal for first use in humans) into an outline designed for a proposal for marketing seems counter-intuitive initially. The intention is that this extra effort will be offset later by the benefits of adding to and revising the original submission. Information in the IND will be revised through amendments to the IND, in which case the old information will be replaced with the new. When the NDA is submitted, numbering in the NDA will be consistent with the IND, thus allowing replacement of old information with new and the addition of new sections that did not originally exist in the IND.

IND content

As for all submissions, an IND is a combination of individual documents (such as a clinical protocol, investigator's brochure, or study report) and text written solely for submission (the introductory statement and general investigational plan). Individual documents should be included 'as is' and not modified specifically to suit the submission. Data from these individual documents are used to write summaries but should not be abbreviated intentionally unless by team consensus.

As with all documents, but particularly for regulatory submissions, which tend to quickly suffer from unrestrained growth, planning ahead is essential to keep within deadline and conserve resources. Figure 2 presents the many uses for well-written synopses in an IND, with implications for economy of time and effort. Nonclinical study report synopses may be used in the nonclinical overview and the investigator's brochure. The synopsis of the clinical protocol is often used in the general investigational plan.

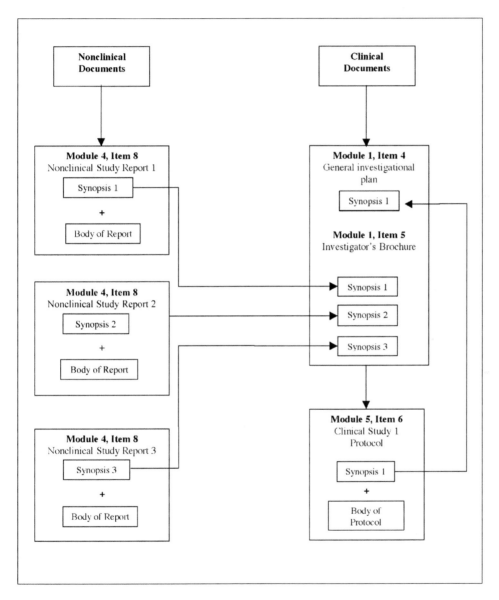

Figure 2. The many uses for well-written synopses in an IND

Item 3: Introductory statement, and Item 4: General investigational plan

The purpose of these items is to provide a brief (very brief: two to three page are suggested) overview of important characteristics of the drug. The information should include the name of the drug and all active ingredients, the drug's pharmacologic class, the structural formula of the drug (if known), the formulation of the dosage form(s) to be used, and the route of administration [3]. In addition, the item should include information on human use:

- Previous use in humans: On occasion, the drug may have studied in another country, for another indication, or have been marketed in another country. A brief sentence or two is appropriate, since this information will be expanded in item 9. Reference to a previous IND, if applicable, is required.
- First planned clinical trial: Provide a brief description including the objectives, study design, population, duration of the study, and outcome assessments. A protocol synopsis (refer to Appendix V) may be used here.
- General investigational plan for the following year: The FDA requires a broad and general idea of clinical investigational plans for the next year, which helps to determine such things as whether or not the nonclinical program and stability testing for the drug are adequate to support clinical research. The investigational plan should be very general because it probably will change greatly during conduct of the first study. A few sentences detailing critical information are all that is required, with an estimate of the number of subjects to be studied.
- Withdrawal from marketing (if applicable): If the drug has been withdrawn from investigation or marketing in any country for any reason related to safety or effectiveness, identify the country(s) where the drug was withdrawn and the reasons for the withdrawal.

Item 5: Investigator's brochure

The investigator's brochure is described in Chapter 7 (Investigator's brochures). The brochure is an individual document and should be included in its entirety. An investigator's brochure represents a summary of all known product information, and is supplied to the investigational sites to ensure safe use in study subjects. No modifications should be made for the sake of the submission unless the modified version is also intended for the investigational sites. Revisions based on updated information, particularly those that might improve safe use, provide clarification, and correct erroneous material are always appropriate, but the resulting updated version must also be supplied to investigational site personnel. Discrepancies between the version submitted to the FDA and the version shipped to the investigational sites implies the potential for misrepresentation of product characteristics.

Side bar: Lessons learned

The reputed enormity of work related to a regulatory submission tends to encourage an initial burst of earnest enthusiasm on the part of inexperienced team members, who feel that 'putting the pedal to the metal' (ie, recruiting the entire department to 'write every-thing down') will be sufficient to meet the timeline with an abundance of documentation (more is always better, after all). The resulting flurry of activity, involving everyone in the vicinity typing furiously 24/7, does truly produce an impressive volume of what might loosely be called information. However, information does not always equal communica-tion, and in this case, often equals overlapping, redundant, poorly organized mountains of material with no focus on key messages. Even earnest enthusiasm fades toward the end of the project, when it becomes clear that much has to be rewritten or just tossed as the deadline looms ever closer and the message of the submission is still buried. The best scenario would be one in which the team takes the time to reassess and pare text down to show the hidden message within. Unfortunately, the restraints of time often mandate that the submission be shipped, in which case the innocent reviewer is the unwilling victim of this excess. Submissions are a marathon, not a sprint. Careful planning and a steady pace are critical to success.

Item 6: Protocols

Clinical protocols are described in Chapter 5. Protocols are individual documents and should be included in their entirety. As with investigator's brochures, the clinical protocol submitted with the IND should be identical to the protocol that is provided to investigational site personnel.

Item 9: Previous human experience with the investigational drug

For most IND submissions, this item will not be applicable. It is possible, however, that the drug has been studied in a country other than the United States, currently is approved for marketing in another country, or has been studied in the past and withdrawn from investigation or marketing. If any experience with administration of the drug to humans exists, this item should contain a summary. The methods of summarizing are similar to those for the investigator's brochure (Chapter 7).

New Drug Application

Since 1938, every drug new to the market in the United States has required submis-sion and approval of a New Drug Application (NDA) before marketing and distribution. Before 1938, drugs were regulated under the Food and Drugs Act of 1906 that was administered by the Bureau of Chemistry. The act regulated product labeling and prohibited interstate transport of mislabeled food or drugs, but it did not address premarket approval. The change in regulation of drug sales was prompted by a therapeutic disaster. In 1937, a Tennessee drug company marketed a sulfa drug to children, Elixir Sulfanilamide. Sulfa drugs, in general, were new to the market and were claimed to hold great promise. Clinical testing of new drugs did not exist as we now know it, and more than 100 people died, many of them children, from a highly toxic chemical analogue of antifreeze used in the elixir's formulation. In response to the public outcry, President Franklin D. Roosevelt signed the Food, Drug, and Cosmetic Act on 25 June 1938 [4].

An NDA is a sponsor's proposal to FDA that testing of the drug has demonstrated sufficient safety, that it is effective for its intended use, that the benefits outweigh the risks, and that the methods used in manufacturing and controls maintain the drug's identity, strength, quality, and purity [5].

The process by which FDA reviews an NDA is presented in Figure 3. Note that upon receipt, FDA makes a determination concerning whether or not the application is 'fileable.' A Refuse to File (RTF) letter is issued for an application that is missing one or more essential components. The letter documents the missing component(s) and informs the applicant that the application will not be filed until it is complete. No further review occurs until the requested data are supplied and the application is found to be acceptable and complete [6]. A RTF letter is a serious outcome, putting years of research, hundreds of millions of dollars, and numerous staff hours at risk. Very often the essential elements that prompt such a letter are not scientific but rather missing sections, page numbers, table of contents, or, in an electronic submission, technical difficulties that preclude review.

Side bar: Lessons learned

The NDA is the submission with the urban legend of requiring an 18-wheel truck for the purpose of transporting hard copies to the FDA. This frightening (and all too often justified) reputation is based not only on the massive quantity of information required but also on the fact that multiple copies of each volume were required. Thankfully, electronic publishing techniques are gradually eliminating hard copy, so eventually the memory of months spent stamping pages with a Bates paginator, standing in a hot, airless room at the copy machine, and binding and tabbing several hundred volumes will fade along with the scars from the paper cuts and the pain from the torn rotator cuff.

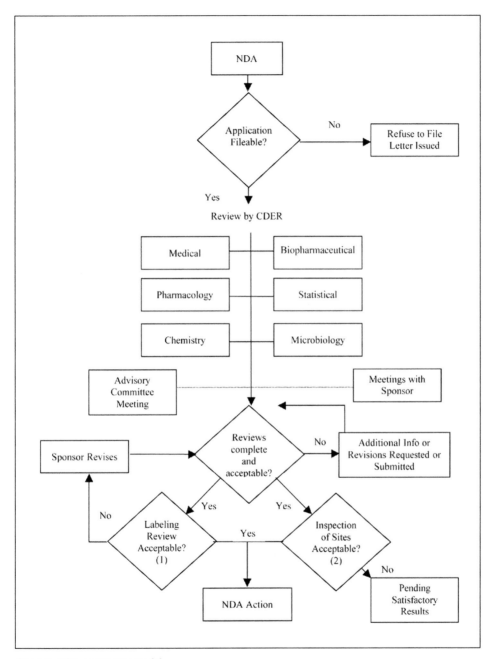

Figure 3. NDA review process [5]
(1) Labeling means instructions for use
(2) Manufacturing sites and sites where significant clinical trials are performed

The clinical sections of an NDA (sections of an NDA are also called items) are presented in Table 3 as defined by the United States CFR. Before the introduction of the CTD format for NDA submissions, NDAs were written using the structure defined on Form FDA 356h and described in Title 21 CFR Part 314.

Table 3. Outline of an NDA as described on Form FDA 365h and in Title 21 CFR Part 314

1. Index
2. Labeling
3. Summary
4. Chemistry
    A. Chemistry, manufacturing, and controls information
    B. Samples
    C. Methods validation package
5. Nonclinical pharmacology and toxicology section
6. Human pharmacokinetics and bioavailability section
7. Clinical microbiology
8. Clinical data section
9. Safety update report
10. Statistical section
11. Case report tabulations
12. Case report forms
13. Patent information of any patent that claims the drug
14. Patent certification with respect to any patient that claims the drug
15. Establishment description
16. Debarment certification
17. Field copy certification
18. User fee cover sheet
19. Financial information
20. Other

Before the CTD, the clinical sections of an NDA were written (loosely, with much interpretation) using the Guideline for the Format and Content of the Clinical and Statistical Sections of an Application [7]. As part of the effort to harmonize global submissions, the NDA is now written to conform to the CTD format. The requirements for an NDA are identical to the requirements in effect before the institution of the CTD, but additional requirements based on the CTD are needed, and the NDA contents have been reordered. Placement of NDA contents into the appropriate sections of the CTD outline is presented in Table 4.

Table 4. Mapping clinical sections of the NDA to CTD format [8]

CFR		CTD outline		
Number	Title	Module	Number	Title
314.50(c)(2)(ii)-(ix)	Item 3: Summaries	2	As needed	Use appropriate sections
314.50(d)(3) 601.2	Item 6: Human pharmacokinetics, bioavailability	5	5.3.1	Use appropriate sections
314.50(d)(4)	Item 7: Clinical microbiology	5	5.3.5.5	Other study reports and related information (Use appropriate sections in microbiology)
314.50(d)(5)(i) 601.2	Item 8: Clinical data	5	5.3	Use appropriate sections
314.50(d)(5)(v) 601.2	Item 8: ISE	5	5.3.4	Integrated report
314.50(d)(5)(vi)(a) 601.2	Item 8: ISS	5	5.3.4	Integrated report
314.50(d)(5)(viii) 601.2	Item 8: Risk/benefit summary	2	2.5	Use appropriate sections

References

1   Center for Drug Evaluation and Research Food and Drug Administration, Department of Health and Human Service, http://www.fda.gov/cder/regulatory/applications/ind_page_1.htm#Introduction, Investigational New Drug (IND) Application Process (Accessed 19 January 2008).
2   Center for Drug Evaluation and Research Food and Drug Administration, Department of Health and Human Service, http://www.fda.gov/cder/handbook/clinhold.htm, Clinical Trial Hold (Accessed 19 January 2008).
3   Center for Drug Evaluation and Research Food and Drug Administration, Department of Health and Human Service, 312.23 Code of Federal Regulations.
4   Center for Drug Evaluation and Research Food and Drug Administration, Department of Health and Human Service, History of the FDA. The 1938 Food, Drug, and Cosmetic Act, http://www.fda.gov/oc/history/historyoffda/section2.html (Accessed 9 March 2008).
5   Center for Drug Evaluation and Research Food and Drug Administration, Department of Health and Human Service, New Drug Development and Review Process, http://www.fda.gov/cder/regulatory/applications/NDA.htm (Accessed 9 March 2008).
6   Center for Drug Evaluation and Research Food and Drug Administration, Department of Health and Human Services, Refuse to File Letter, http://www.fda.gov/cder/handbook/refusegn.htm (Accessed 9 March 2008).
7   Center for Drug Evaluation and Research Food and Drug Administration, Department of Health and Human Services, Guideline for the Format and Content of the Clinical and Statistical Sections of an Application, July 1988.
8   Center for Drug Evaluation and Research Food and Drug Administration, Department of Health and Human Services, Comprehensive Table of Headings and Hierarchy, http://www.fda.gov/Cder/regulatory/ersr/5640CTOC-v1.2.pdf (Accessed 29 February 2008).

Appendices

Appendix 1

Regulatory review checklists

Protocol Quality Control Checklist

Title of Document:	
Protocol Number:	
Signature of Reviewer:	
Date:	

Checked box indicates that the item has been reviewed.

1. Title Page Information
   - ☐ Title has the correct investigational product name and reflects the study design
   - ☐ Protocol number consistent with internal policies
   - ☐ Sponsor name and address correct
   - ☐ Indication correct
   - ☐ Date reflects the final protocol or amendments (if applicable)
   - ☐ GCP compliance statement present

2. Tables of Contents
   - ☐ All section headings, subheadings, and appendices present
   - ☐ Title format consistent
   - ☐ Heading, subheading, table, figure, and appendix numbers correct
   - ☐ Page numbers for all headings, subheadings, tables, figures, and appendices correct
   - ☐ No widows or orphans

3. List of Tables
   - ☐ All tables appear in list
   - ☐ Title format consistent
   - ☐ Table numbers correct
   - ☐ Page numbers of tables correct
   - ☐ No widows or orphans

4. List of Figures
   - ☐ All figures appear in list
   - ☐ Title consistently formatted
   - ☐ Figure numbers correct
   - ☐ Page numbers of figures correct
   - ☐ No widows or orphans

5. Tables
   - ☐ Numbers (or other table entries) match source table(s)
   - ☐ No data missing unless explained clearly
   - ☐ Source of data correctly cited with each table
   - ☐ Table title reflects the content and is consistent with the source table
   - ☐ Title appears on the same page as the table
   - ☐ All footnotes in the table defined below the table

☐ Column and row headers correct
☐ Overall table formatting correct

6. Figures
    ☐ Figure title reflects content and is consistent with source figure
    ☐ Figure title appears on the same page as the figure
    ☐ All legends, units, and labels on axes correct

7. Text
    ☐ Verb tense = future
    ☐ Text in the synopsis agrees with the text in the body of the protocol
    ☐ Cross references to tables, figures, and/or appendices correct
    ☐ Numbers in the text match text tables (or source tables, if no text table)
    ☐ Headings, subheadings, table and figure titles format consistent in text
    ☐ No typographical errors or misspellings

8. Abbreviations
    ☐ Defined in text, first in the synopsis and again in the body of the protocol
    ☐ Used consistently throughout the text
    ☐ All abbreviations defined
    ☐ Ordered alphabetically
    ☐ Format consistent

9. Headers and Footers
    ☐ Appear on each page
    ☐ Date consistent
    ☐ Page numbers consecutive
    ☐ Total number of pages consistent
    ☐ Appropriately formatted for orientation

10. References
    ☐ Numbered consecutively
    ☐ Format consistent
    ☐ Spacing between references consistent
    ☐ Each reference cited in text
    ☐ References listed in order cited in text

11. Appendices
    ☐ Numbering consecutive
    ☐ Complete, or if abbreviated, explained

Clinical Study Report Quality Control Checklist

Title of Document:	
Protocol Number:	
Signature of Reviewer:	
Date:	

Checked box indicates item has been reviewed.

1. Title Page Information
   - ☐ Title and study report number match protocol
   - ☐ Sponsor names correct
   - ☐ Date of study start/stop matches synopsis dates
   - ☐ Date matches synopsis date of report

2. Tables of Contents
   - ☐ All section headings, subheadings, and appendices present
   - ☐ Title format consistent
   - ☐ Heading, subheading, table, figure, and appendix numbers correct
   - ☐ Page numbers for all headings, subheadings, tables, figures, and appendices correct
   - ☐ No widows or orphans

3. List of Tables
   - ☐ All tables appear in list
   - ☐ Titles format consistent
   - ☐ Table numbers correct
   - ☐ Page numbers of tables correct
   - ☐ No widows or orphans

4. List of Figures (LOF)
   - ☐ All figures appear in list
   - ☐ Titles format consistent
   - ☐ Figure numbers correct
   - ☐ Page numbers of figures correct
   - ☐ No widows or orphans

5. Tables
   - ☐ Numbers (or other table entries) match source table(s)
   - ☐ No data missing
   - ☐ Source of data correctly cited with each table
   - ☐ Table title reflects the content and is consistent with the source table
   - ☐ Title appears on the same page as the table
   - ☐ All footnotes in the table defined below the table
   - ☐ Column and row headers correct
   - ☐ Overall table formatting correct

6. Figures
   - ☐ Figure title reflects content and is consistent with source table
   - ☐ Figure title appears on the same page as the figure
   - ☐ All legends, units, and labels on axes correct

7. Text
    ☐ Verb tense = past intention for methods sections, past tense for results
    ☐ Text in the synopsis agrees with the text in the body of the report
    ☐ Cross references to tables, figures, and/or appendices correct
    ☐ Numbers in the text match text tables (or source tables if no text table)
    ☐ Headings, subheadings, table and figure titles format consistent in text
    ☐ No typographical errors or misspellings

8. Abbreviations
    ☐ Defined in text, first in the synopsis and again then in the body of the report
    ☐ Used consistently throughout the text
    ☐ All abbreviations defined
    ☐ Ordered alphabetically
    ☐ Formatting consistent

9. Headers and Footers
    ☐ Appear on each page
    ☐ Date consistent
    ☐ Page numbers consecutive
    ☐ Total number of pages consistent

10. References
    ☐ Numbered consecutively
    ☐ Format consistent
    ☐ Spacing between references consistent
    ☐ Each reference cited in text
    ☐ References listed in order cited in text

11. Appendices
    ☐ Numbered consecutively

Appendix II

Sample clinical protocol outline

Title page
1   Synopsis

2   Study Schedule/Schema

3   Table of Contents

4   List of Abbreviations

5   Background and Rationale
5.1   Disease
5.2   Background on Investigational Product
5.3   Rationale and Hypothesis

6   Objectives
6.1   Primary Objective
6.2   Secondary Objectives

7   Experimental Plan
7.1   Study Design
7.2   Number of Centers and Subjects
7.3   Estimated Study Duration

8   Subject Selection
8.1   Number of Subjects
8.2   Inclusion Criteria
8.3   Exclusion Criteria
8.4   Withdrawal

9   Schedule of Assessments and Procedures
9.1   Screening
9.2   Enrollment, Randomization
9.3   Week 1: Baseline Evaluations
9.4   Weeks 3, 5, 7
9.5   Week 8: End of Study

10   Investigational Product
10.1   Investigational Drug
10.2   Control Drug
10.3   Randomization to Treatment
10.4   Dosing and Administration
       10.4.1   Dose Escalation, Dose Adjustments, Stopping Rules
       10.4.2   Blinding
       10.4.3   Packaging and Labelling
       10.4.4   Storage
       10.4.5   Accountability
10.5   Concomitant Medications

Appendix III

Sample clinical protocol title page

Title:	A Phase 3, Multicenter, Prospective, Randomized, Controlled Comparison of Panacea With Marketed Product in Men With Androgenetic Alopecia
Protocol Number:	CLN 001 07
Test Drug:	panacea acetate
Indication:	For use in men with androgenetic alopecia
Study Design	Phase 3, multicenter, prospective, randomized, controlled comparison of panacea with marketed product
Sponsor Name, Address, and Telephone Number:	New Therapeutics, Inc. 1212 Therapy Parkway Big City, NY 10001 555.123.4567
Sponsor's Representative:	John Doe, MD, PhD Vice President of Clinical Research and Development Office: 555.123.1234 Fax: 555.123.0987
Monitor Name, Address, and Telephone Number:	Jane Smith, MD Director, Medical Research 1212 Therapy Parkway Big City, NY 10001 Office: 555.123.0725 Fax: 555.123.5556
Compliance:	The study will be conducted in accordance with standards of Good Clinical Practice, as defined by the International Conference on Harmonisation and all applicable federal and local regulations.
Date of Protocol:	7 August 2007

Appendix IV

Sample clinical protocol signature page

I have read this clinical protocol and confirm that to the best of my knowledge it accurately describes the design and conduct of the study titled 'A Phase 3, Multicenter, Prospective, Randomized, Controlled Comparison of Panacea With Marketed Product in Men With Androgenetic Alopecia.'

George Brown, MD, PhD   Date

_____   _____

Name (please print)

Appendix V

Sample clinical protocol synopsis

Name of Sponsor/Company: New Therapeutics, Inc.
Name of product: Panacea
Title of study: A Phase 3, Multicenter, Prospective, Randomized, Controlled Comparison of Panacea with Marketed Product in Men with Androgenetic Alopecia.
Objectives: Primary: To assess the efficacy of panacea when used in men with androgenetic alopecia. Secondary: To collect safety data for the oral formulation.
Methodology: This study is a phase 3, multicenter, prospective, randomized, controlled comparison of panacea with standard therapy (marketed product) in men with androgenetic alopecia. Men will be screened within 2 weeks of starting treatment. Screening examinations will consist of a physical examination, height, weight, vital signs, an electrocardiogram (ECG), medical history, current medications, and informed consent. Sampling for routine hematology, chemistry, and urinalysis will be performed. Hair counts will be conducted on a 1-inch circle of scalp, and the area will be photographed. Eligible men will be randomly assigned in a 1:1 ratio to a treatment group with panacea or marketed product and receive a 2-week supply of blister-packed study treatment. Subjects will return on weeks 3, 5, and 7 for a hair count, for photographs, and an Investigator Assessment of Hair Growth (IAHG), and to complete a Patient Assessment of Hair Growth (PAHG). During each of these visits the subjects will receive a 2-week supply of investigational drug, and will be questioned about concomitant medications and adverse events. All subjects will return on week 8 for a final study visit. This visit will include a physical examination, weight, vital signs, an ECG, current medications, and questioning concerning adverse events. Sampling for routine hematology, chemistry, and urinalysis will be performed. Hair counts will be conducted on a 1-inch circle of scalp, and the area will be photographed.
Number of subjects planned: 500 men at 30 investigational sites in the United States.
Diagnosis and main criteria for inclusion: Subjects will be men, aged 18 to 50 years, with mild to moderate hair loss of the vertex and anterior midscalp area, and no history or current laboratory evidence of renal, hepatic, or cardiac condition that could potentially put them at risk while using this test product.
Test product, dose, and mode of administration (proposed): Panacea and marketed product will be supplied in blister packs of 14 capsules each (a 2-week supply). Matched placebo capsules will also be supplied. Investigational drug is to be administered orally once a day for 8 weeks.
Duration of treatment: 8 weeks

Criteria for evaluation:
Efficacy:
  • Hair counts
  • Photographs of 1-inch circle of scalp
  • PAHG
  • IAHG

Safety:
- Physical examination
- Weight
- Vital signs
- ECG
- Laboratory values
- Adverse events

Statistical analysis: Demographic and baseline characteristics will be summarized using descriptive statistics. Hair counts and the IAHG and PAHG will be analyzed using analysis of variance. Photographs will be evaluated by an independent panel blinded to study treatment.

Appendix VI

Sample list of abbreviations

Abbreviation	Definition
AE	Adverse event
CBC	Complete blood count
CT	Computed tomography
ECG	Electrocardiogram
IRB	Institutional Review Board
MRI	Magnetic resonance imaging
PK	Pharmacokinetics
RBC	Red blood cells
WBC	White blood cells

Appendix VII

Sample protocol amendment

Protocol amendment 1

Title:	A Phase 3, Multicenter, Prospective, Randomized, Controlled Comparison of Panacea With Marketed Product in Men With Androgenetic Alopecia
Protocol Number:	CLN 001 07
Test Drug:	panacea
Indication:	For use in men with androgenetic alopecia
Phase:	3
Sponsor Name, Address, and Telephone Number:	New Therapeutics, Inc. 1212 Therapy Parkway Big City, NY 10001 555.123.4567
Sponsor's Representative:	John Doe, MD, PhD Vice President of Clinical Research and Development Office: 555.123.1234 Fax: 555.123.0987
Monitor Name, Address, and Telephone Number:	Jane Smith, MD Director, Medical Research 1212 Therapy Parkway Big City, NY 10001 Office: 555.123.0725 Fax: 555.123.5556
Date of Protocol Amendment:	17 October 2007

Signature Page

I have read this clinical protocol amendment and confirm that to the best of my knowledge it accurately describes revisions to the protocol titled 'A Phase 3, Multicenter, Prospective, Randomized, Controlled Comparison of Panacea With Marketed Product in Men With Androgenetic Alopecia.'

_____   _____

Signature   Date

_____

Name (please print)

Purpose of Amendment

The purpose of this amendment is to make the following revisions to the clinical protocol:

- Name change from panacetex acetate to panacea acetate to distinguish from another investigational product
- Inclusion criteria will now include women and exclude those with urinary conditions that could put the subject at risk
- Revise labeling on the vials to include the amount of panacea acetate

Changes to the 7 August 2007 Protocol

New text is presented in **Bold** typeface, deleted text is presented with a ~~Strikethrough~~

Product Name Change
    From: panacetex acetate
    To: panacea acetate

Synopsis, Diagnosis and Main Criteria for Inclusion:
Formerly read:
Subjects will be men, aged between 18 and 50 years, with mild to moderate hair loss of the vertex and anterior midscalp area and no history or current laboratory evidence of renal, hepatic, or cardiac condition that could potentially put the subject at risk while using this product.

Now reads:
Subjects will be men **and women**, aged **more than 18 years**, with mild to moderate hair loss of the vertex and anterior midscalp area and no history or current laboratory evidence of renal, hepatic, **urinary**, or cardiac condition that could potentially put the subject at risk while using this product.

7.2 Inclusion Criteria:
Formerly read:
Subjects will be men, aged 18 to 50 years, with mild to moderate hair loss of the vertex and anterior midscalp area and no history or current laboratory evidence of renal, hepatic, or cardiac condition that could potentially put the subject at risk while using this product.

Now reads:
Subjects will be men **and women**, aged **more than 18 years**, with mild to moderate hair loss of the vertex and anterior midscalp area and no history or current laboratory evidence of renal, hepatic, **urinary**, or cardiac condition that could potentially put the subject at risk while using this product.

10.2 Packaging and Labeling:

Formerly read:

panacetex acetate Lot # XXXX Contents: 2 mg panacetex acetate[CE1] Store at 2° to 8°C (35.6° to 46.4°F) Manufacturer - New Therapeutics, Inc., Big City, NY Caution: New Drug – Limited by Federal Law to Investigational Use

Now reads:

panacea acetate ~~panacetex acetate~~ Lot # XXXX Contents: 2 mg panacea acetate ~~panacetex acetate~~ Store at 2° to 8°C (35.6° to 46.4°F) Manufacturer - New Therapeutics, Inc., Big City, NY Caution: New Drug – Limited by Federal Law to Investigational Use

Appendix VIII

Sample clinical study report title page

Title:	A Phase 3, Multicenter, Prospective, Randomized, Controlled Comparison of Panacea With Marketed Product in Men With Androgenetic Alopecia
Investigational Product:	Panacea acetate
Indication:	For use in men with androgenetic alopecia
Methods:	Phase 3, multicenter, prospective, randomized, controlled comparison of panacea with marketed product
Sponsor Name and Address:	New Therapeutics, Inc. 1212 Therapy Parkway Big City, NY 10001 555.123.4567
Protocol Identification:	CLN 001 07
Development Phase:	3
Study Initiation Date:	29 August 2005
Study Completion Date:	19 February 2007
Principal Investigators:	George Brown MD PhD New York Medical Center West 33rd Street New York, NY
Compliance Statement:	The study will be conducted in accordance with standards of Good Clinical Practice, as defined by the International Conference on Harmonisation and all applicable federal and local regulations.
Company Sponsor/Representative:	John Doe, MD, PhD Vice President of Clinical Research and Development Office: 555.123.1234 Fax: 555.123.0987
Date of Report:	29 March 2008

Appendix IX

Sample clinical study report synopsis

Name of Sponsor/Company: New Therapeutics, Inc.	Individual Study Table Referring to Part of the Dossier	*(For National Authority Use)*
Name of Finished Product: Panacea	Volume: Page:	
Name of Active Ingredient: Panacea acetate		

Title of STUDY:
A Phase 3, Multicenter, Prospective, Randomized, Controlled Comparison of Panacea With Marketed Product in Men With Androgenetic Alopecia

Study No.: CLN 001 07

Principal Investigator:
George Brown MD PhD, New York Medical Center, West 33rd Street, New York, NY

Publication (reference): None

Studied Period: Study initiation date: 29 August 2005 Study completion date: 19 February 2007	Phase of development: 3

Objectives:
Primary: To assess the efficacy of panacea when used in men with androgenetic alopecia after 8 weeks of treatment Secondary: To collect safety data for the oral formulation

Methodology:
This study was a phase 3, multicenter, prospective, randomized, controlled comparison of panacea with standard therapy (marketed product) in men with androgenetic alopecia. Men were to be screened within 2 weeks of starting treatment. Screening examinations were to consist of a physical examination, height, weight, vital signs, electrocardiogram (ECG), medical history, current medications, and informed consent. Sampling for routine hematology, chemistry, and urinalysis was to be performed. Hair counts were to be conducted on a 1-inch circle of scalp, and the area was to be photographed. Eligible men were randomly assigned in a 1:1 ratio to treatment group with panacea or marketed product and received a 2-week supply of blister-packed study treatment. Subjects returned on weeks 3, 5, and 7 for a hair count, for photographs, an Investigator Assessment of Hair Growth (IAHG), and to complete a Patient Assessment of Hair Growth (PAHG). During each of these visits, the subjects received a 2-week supply of study drug and were questioned about concomitant medications and adverse events. All subjects returned on week 8 for a final study visit. This visit included a physical examination, weight, vital signs, ECG, current medications, and questioning concerning adverse events. Sampling for routine hematology, chemistry, and urinalysis was performed. Hair counts were conducted on a 1-inch circle of scalp, and the areas were photographed.

Name of Sponsor/Company: New Therapeutics, Inc.	Individual Study Table Referring to Part of the Dossier	(*For National Authority Use*)
Name of Finished Product: Panacea	Volume: Page:	
Name of Active Ingredient: Panacea acetate		

Number of Subjects Planned:

500 men at 30 investigational sites in the United States were planned. Four hundred thirty-eight subjects were enrolled, 216 in the test group and 222 in the control group.

Diagnosis and Main Criteria for Inclusion:

Subjects were men, aged 18 to 50 years, with mild to moderate hair loss of the vertex and anterior midscalp area and no history or current laboratory evidence of renal, hepatic, or cardiac condition that could potentially put them at risk while using this test product.

Test Product, Dose, and Mode of Administration (Proposed):

Panacea and marketed product were supplied in blister packs of 14 capsules each (a 2-week supply). Matched placebo capsules also were supplied. Study drug was to be administered orally once a day for 8 weeks.

Duration of Treatment: 8 weeks

Criteria for Evaluation:

Efficacy:

- Hair counts
- Photographs of 1-inch circle of scalp
- Investigator Assessment of Hair Growth (IAHG)
- Patient Assessment of Hair Growth (PAHG)

Safety:

- Physical examination
- Weight
- Vital signs
- ECG
- Laboratory values
- Adverse events

Demographic and baseline characteristics were summarized using descriptive statistics. Hair counts, and the IAHG and PAHG were analyzed using analysis of variance. Photographs were evaluated by an independent panel. Determination of efficacy was made 12 weeks post-initiation of treatment.

Name of Sponsor/Company: New Therapeutics, Inc.	Individual Study Table Referring to Part of the Dossier	(*For National Authority Use*)
Name of Finished Product: Panacea	Volume: Page:	
Name of Active Ingredient: Panacea acetate		

Results

Disposition:

Four hundred thirty-eight subjects were enrolled, 216 in the test group and 222 in the control group. All except 6 subjects received study medication, 4 in the test group and 2 in the control group. Four hundred fourteen subjects completed the study. Reasons for premature withdrawal were primarily loss to follow-up. Subject 02 in the control group died of a cardiac arrest. Disposition is presented in Table 1.

Table 1. Disposition

Disposition	Panacea	Control	Total
Enrolled/randomized	216	222	438
Treated	212	220	432
Not treated	4	2	6
Completed study	206	208	414
Withdrawn prematurely	6	12	18
Adverse event	2	3	5
Death	0	1	1
Lost to follow up	4	8	12

All subjects treated with study medication were included in the safety dataset, which was used for safety evaluations. The intent-to-treat dataset included 6 subjects who did not receive study medication. The per protocol (PP) dataset included subjects who completed the study. Efficacy analyses were performed on both the ITT and PP datasets. Datasets used for analysis are presented in Table 2.

Name of Sponsor/Company: New Therapeutics, Inc.	Individual Study Table Referring to Part of the Dossier	(*For National Authority Use*)
Name of Finished Product: Panacea	Volume: Page:	
Name of Active Ingredient: Panacea acetate		

Table 2. Datasets Used for Analysis

Dataset	Panacea N = 216	Control N = 222	Total N = 438
Enrolled/randomized	216	222	438
Treated	212	220	432
Not treated	4	2	6
Intent-to-treat dataset	216	222	438
Safety dataset	212	220	432
Per protocol dataset	206	208	414

Demographics and Baseline Characteristics:

The groups were comparable with respect to all demographic and baseline characteristics. All subjects were men, mean ± standard deviation age was 62 ± 6.2, 80% were white, and none had a history of a serious medical condition.

Baseline mean hair counts, PAHG, and IAGH were not statistically significantly different. Photographs, evaluated by an independent panel, did not suggest differences in alopecia at baseline (Table 3).

Table 3. Alopecia Assessments at Baseline

Outcome	Panacea N = 216 Mean	Control N = 222 Mean	P Value
Hair count (n/1 inch circle)	5	7	0.8654
PAHG (score 1-10)	3	4	0.7598
IAGH (score 1-10)	4	4	0.9234
Photographs (score 1-10)	2	3	0.8562

Efficacy:

The primary endpoint of the study was met: Subjects who were treated with panacea showed statistically significantly greater hair counts and PAHG and IAHG scores 8 weeks after initiation of treatment ($P<0.0001$ for all alopecia assessments; Table 4). Mean hair counts increased in the panacea-treated group from 5 to 45 per 1-inch circle compared with the control group in which hair counts increased from

Name of Sponsor/Company: New Therapeutics, Inc.	Individual Study Table Referring to Part of the Dossier	(*For National* *Authority Use*)
Name of Finished Product: Panacea	Volume: Page:	
Name of Active Ingredient: Panacea acetate		

7 to 9 per 1-inch circle. PAHG and IAHG also increased in the panacea-treated group, although investigator assessment of improvement exceeded that of subject assessment.

Photographs showed a statistically significant difference between groups (*P*=0.0443) despite the inherent subjectivity of this type of assessment.

Table 4. Alopecia Assessments at 8 Weeks

Outcome	Panacea N = 216 Mean	Control N = 222 Mean	*P* Value
Hair count (n per 1 inch circle)	45	9	<0.0001
PAHG (score 1-10)	7	5	<0.0001
IAGH (score 1-10)	8	4	<0.0001
Photographs (score 1-10)	5	3	0.0443

Safety:

Extent of Exposure:

Subjects in the panacea group received 2 mg of panacea acetate daily for 8 weeks, for a total of 448 mg.

Adverse Events:

Three hundred forty-two adverse events occurred in 98 subjects during the study period. With the exception of the death of subject 02 (cardiac arrest) none of the adverse events were considered to be serious or life threatening, and all but 5 were mild to moderate in severity. Five subjects (2 in the panacea group and 3 in the control group) withdrew from the study due to severe adverse events. These events were gastric irritation occurring within 20 minutes after dosing. After discontinuation of study medication, all gastric irritation resolved without sequelae.

Conclusions:

Panacea 2 mg orally daily for 8 weeks successfully increased hair growth in men with androgenetic alopecia in this multicenter, prospective, randomized, controlled, comparison with standard therapy. Hair growth was increased by all assessments [hair counts and PAHG and IAHG scores (*P*<0.0001) for all differences between test article and control] after 8 weeks of dosing. Panacea appears to be safe and well tolerated. Transient gastric irritation occurred after dosing, but frequency was similar for the study groups.

Date of the report: 29 March 2008

Appendix X

Clinical study report outline: ICH E3 and suggested versions

ICH E3 Outline

1.   TITLE PAGE
2.   SYNOPSIS
3.   TABLE OF CONTENTS FOR THE INDIVIDUAL CLINICAL STUDY REPORT
4.   LIST OF ABBREVIATIONS AND DEFINITIONS OF TERMS
5.   ETHICS
5.1   Independent Ethics Committee (IEC) or Institutional Review Board (IRB)
5.2   Ethical Conduct of the Study
5.3   Subject Information and Consent
6.   INVESTIGATORS AND STUDY ADMINISTRATIVE STRUCTURE
7.   INTRODUCTION
8.   STUDY OBJECTIVES
9.   INVESTIGATIONAL PLAN
    9.1   Overall Study Design and Plan: Description
    9.2   Discussion of Study Design, Including the Choice of Control Groups
    9.3   Selection of Study Population
        9.3.1   Inclusion Criteria
        9.3.2   Exclusion Criteria
        9.3.3   Removal of Subjects From Therapy or Assessment
    9.4   Treatments
        9.4.1   Treatments Administered
        9.4.2   Identity of Investigational Products(s)
        9.4.3   Method of Assigning Subjects to Treatment Groups
        9.4.4   Selection of Doses in the Study
        9.4.5   Selection and Timing of Dose for Each Subject
        9.4.6   Blinding
        9.4.7   Prior and Concomitant Therapy
        9.4.8   Treatment Compliance
    9.5   Efficacy and Safety Variables
        9.5.1   Efficacy and Safety Measurements Assessed and Flow Chart
        9.5.2   Appropriateness of Measurements
        9.5.3   Primary Efficacy Variable(s)
        9.5.4   Drug Concentration Measurements
    9.6   Data Quality Assurance
    9.7   Statistical Methods Planned in the Protocol and Determination of Sample Size
        9.7.1   Statistical and Analytical Plans
        9.7.2   Determination of Sample Size
    9.8   Changes in the Conduct of the Study or Planned Analyses
10. STUDY SUBJECTS
    10.1   Disposition of Subjects
    10.2   Protocol Deviations
11. EFFICACY EVALUATION
    11.1   Data Sets Analyzed
    11.2   Demographic and Other Baseline Characteristics
    11.3   Measurements of Treatment Compliance
    11.4   Efficacy Results and Tabulations of Individual Subject Data

16.1.2   Sample case report form (unique pages only)
16.1.3   List of IECs or IRBs (plus the name of the committee chair if required by the regulatory authority) and representative written information for subject and sample consent forms
16.1.4   List and description of investigators and other important participants in the study, including brief (one page) CVs or equivalent summaries of training and experience relevant to the performance of the clinical study
16.1.5   Signatures of principal or coordinating investigator(s) or sponsor's responsible medical officer, depending on the regulatory authority's requirement
16.1.6   Listing of subjects receiving test drug(s)/investigational product(s) from specific batches, where more than one batch was used
16.1.7   Randomization scheme and codes (subject identification and treatment assigned)
16.1.8   Audit certificates (if available)
16.1.9   Documentation of statistical methods
16.1.10   Documentation of inter-laboratory standardization methods and quality assurance procedures if used
16.1.11   Publications based on the study
16.1.12   Important publications referenced in the report
16.2   Subject Data Listings
16.2.1   Discontinued subjects
16.2.2   Protocol deviations
16.2.3   Subjects excluded from the efficacy analysis
16.2.4   Demographic data
16.2.5   Compliance and/or drug concentration data (if available)
16.2.6   Individual efficacy response data
16.2.7   Adverse event listings (each subject)
16.2.8   Listing of individual laboratory measurements by subject, when required by regulatory authorities
16.3   Case Report Forms (CRFs)
16.3.1   CRFs for deaths, other serious adverse events, and withdrawals for adverse events
16.3.2   Other CRFs submitted
16.4   Individual Subject Data Listings

Suggested Outline

1.   TITLE PAGE
2.   SYNOPSIS
3.   TABLE OF CONTENTS FOR THE INDIVIDUAL CLINICAL STUDY REPORT
4.   LIST OF ABBREVIATIONS AND DEFINITIONS OF TERMS
5.   ETHICS
    5.1   Independent Ethics Committee (IEC) or Institutional Review Board (IRB)
    5.2   Ethical Conduct of the Study
    5.3   Subject Information and Consent
6.   INVESTIGATORS AND STUDY ADMINISTRATIVE STRUCTURE
7.   INTRODUCTION
8.   STUDY OBJECTIVES
9.   INVESTIGATIONAL PLAN
    9.1   Overall Study Design and Plan: Description
    9.2   Discussion of Study Design, Including the Choice of Control Groups
    9.3   Selection of Study Population
        9.3.1   Inclusion Criteria
        9.3.2   Exclusion Criteria
        9.3.3   Removal of Subjects From Therapy or Assessment
    9.4   Treatments
        9.4.1   Treatments Administered
        9.4.2   Identity of Investigational Products(s)
        9.4.3   Method of Assigning Subjects to Treatment Groups
        9.4.4   Selection of Doses in the Study
        9.4.5   Selection and Timing of Dose for Each Subject
        9.4.6   Blinding
        9.4.7   Prior and Concomitant Therapy
        9.4.8   Treatment Compliance
    9.5   Efficacy and Safety Variables
        9.5.1   Efficacy and Safety Measurements Assessed and Flow Chart
        9.5.2   Appropriateness of Measurements
        9.5.3   Primary Efficacy Variable(s)
        9.5.4   Drug Concentration Measurements
    9.6   Data Quality Assurance
    9.7   Statistical Methods Planned in the Protocol and Determination of Sample Size
        9.7.1   Statistical and Analytical Plans
        9.7.2   Determination of Sample Size
    9.8   Changes in the Conduct of the Study or Planned Analyses
10.   STUDY SUBJECTS
    10.1   Disposition of Subjects
    10.2   Data Sets Analyzed
    10.3   Demographic and Other Baseline Characteristics
    10.4   Measurements of Treatment Compliance
    10.5   Protocol Deviations
11.   EFFICACY EVALUATION
    11.1   Primary Efficacy Outcome
    11.2   Secondary Efficacy Outcomes
    11.3   Pharmacokinetics/Pharmacodynamics
    11.4   Pharmacoeconomics
    11.5   Efficacy Conclusions
12.   SAFETY EVALUATION
    12.1   Extent of Exposure
    12.2   Adverse Events
        12.2.1   Overall Adverse Events

Appendix XI

Sample investigator's brochure outline

Title Page (With Confidentiality Statement)
Signature Page
List of Abbreviations
1.   Table of Contents
2.   Summary
  2.1 Description of the Investigational Product
  2.2 Nonclinical Summary
  2.3 Clinical Summary
3.   Introduction
  3.1 Disease
  3.2 Investigational Product
  3.3 Rationale for Clinical Program
4.   Physical, Chemical, and Pharmaceutical Properties and Formulation
  4.1 Description
  4.2 Structure and Physical Properties
  4.3 Dosage Form
  4.4 Handling and Preparation
  4.5 Administration
  4.6 Supply and Storage
5.   Nonclinical Studies
  5.1 Overview with Tabular Summary
  5.2 Nonclinical Pharmacology
    5.2.1 Pharmacology in Rats With Intravenous Dosing
    5.2.2 Pharmacology in Dogs With Intravenous Dosing
    5.2.3 Pharmacology in Rats With Oral Dosing
    5.2.4 Pharmacology in Dogs With Oral Dosing
  5.3 Pharmacokinetics and Product Metabolism in Animals
    5.3.1 Mechanism of Action
    5.3.2 Pharmacokinetics in Rats After Intravenous Dosing
    5.3.3 Pharmacokinetics in Rats After Oral Dosing
    5.3.4 Pharmacokinetics in Dogs After Oral Dosing
  5.4 Toxicology
    5.4.1 Acute Toxicity in Rats
    5.4.2 Acute Toxicity in Dogs
    5.4.3 Chronic Toxicity in Rats
    5.4.4 Chronic Toxicity in Dogs
    5.4.5 Reproductive Toxicity
6.   Effects in Humans
  6.1 Overview with Tabular Summary
  6.2 Pharmacokinetics and Product Metabolism in Humans
    6.2.1 Pharmacokinetics After Intravenous Dosing
    6.2.2 Pharmacokinetics After Oral Dosing
  6.2 Efficacy
    6.2.1 Efficacy With Intravenous Dosing
    6.2.2 Efficacy With Oral Dosing
  6.3 Safety
    6.3.1 Exposure
    6.3.2 Adverse Events

Appendix XII

Investigational medicinal products dossier previous human experience outline

2.3.1.   Clinical Pharmacology
      2.3.1.1. Brief Summary
      2.3.1.2. Mechanism of Primary Action
      2.3.1.3. Secondary Pharmacological Effects
      2.3.1.4. Pharmacodynamic Interactions
2.3.2.   Clinical Pharmacokinetics
      2.3.2.1. Brief Summary
      2.3.2.2. Absorption
      2.3.2.3. Distribution
      2.3.2.4. Elimination
      2.3.2.5. Pharmacokinetics of Active Metabolites
      2.3.2.6. Plasma Concentration-Effect Relationship
      2.3.2.7. Dose and Time Dependencies
      2.3.2.8. Special Patient Populations
      2.3.2.9. Interactions
2.3.3.   Human Exposure
      2.3.3.1. Brief Summary
      2.3.3.2 Overview of Safety and Efficacy
      2.3.3.3. Healthy Subject Studies
      2.3.3.4. Patient Studies
      2.3.3.5. Previous Human Experience
      2.3.4.   Benefits and Risks Assessment
4.   Appendices

Appendix XIII

Sample informed consent form

Title of Study: Comparison of Tumorigen and Miracell in the Treatment of Patients With
                             Advanced Breast Cancer
Principal Investigator: Dr Marion Good
Institute: Department of Hematology and Oncology, Big University
Sponsor: BioPharma, Ltd

Introduction

I am Dr Marion Good from the Department of Hematology and Oncology, Big University, and I am
doing research with BioPharma, Ltd on the treatment of advanced breast cancer. A new drug, tumorigen, is being recommended for the treatment of patients, men and women, with advanced breast cancer.
We want to know whether tumorigen is as good as, or better than, the commonly used drug, miracell, for
treating advanced breast cancer. Since you have been diagnosed with advanced breast cancer, we invite
you to join this research study.

Background Information

Breast cancer is a common disease. A new drug known as tumorigen is thought to be effective in
treating advanced breast cancer, but we do not have enough evidence that it is as good as other drugs
currently used for treating advanced breast cancer.

Purpose of This Research Study

The purpose of study is to find out whether the new drug, tumorigen, is as good as, or better than, other
drugs used for treating advanced breast cancer.

Procedures

In this study, all patients aged 18 to 50 years of age, coming to the clinic with advanced breast cancer will
be registered. The patients will be divided randomly into 2 groups by a computer draw. One group will
get the new drug (tumorigen), and the other group will get the commonly used drug (miracell). Neither
the doctor nor the patient will know which drug the patient is getting for treatment of his/her disease. A
record will be kept during the treatment and will also be used to record other symptoms, including any
side effects. Other necessary treatments will also be provided to you, if needed.

Possible Risks or Benefits

No significant side effects have been reported for tumorigen; however, some patients may feel nausea or
may vomit. Drawing of blood may cause some discomfort or blue discoloration at the site of the blood
draw.
You will not receive any direct financial or other benefit for participating in the study. However, all
examinations, laboratory tests, and other diagnostic tests will be done free of cost to you, and the drugs
(tumorigen or miracell) will be provided for free during the study. Treatment of any side effect will be
provided for free. The sponsor of the study will pay for the drugs, investigations, and treatment of any
side effects related to the study drugs.

Right of Refusal to Participate and Withdrawal

You are free to choose to participate in the study. You may refuse to participate without losing any benefit that you are otherwise entitled to. You may also withdraw at any time from the study without any
effect on the management of your cancer or any loss of benefits that you are otherwise entitled to. You
may also refuse to answer some or all questions if you do not feel comfortable with them.

Confidentiality
The information you provide will remain confidential. No one except the principal investigator will be able to look at your data. Your name and identity will not be revealed at any time. The data (with no patient identification) may be seen by the Ethical Review Committee, clinical development staff at BioPharma, Ltd, and governmental agencies responsible for granting permission to sell the drug once it is found to be safe and helpful. The data may be published in a journal and elsewhere, but your name and identity will not be used.

Available Sources of Information
If you have any other questions, you may contact the principal investigator (Dr Marion Good) at 555.555.5555.

AUTHORIZATION

I have read and understand this consent form, and I volunteer to participate in this research study. I understand that I will receive a copy of this form. I voluntarily choose to participate, but I understand that my consent does not take away any legal rights in the case of negligence or other legal fault of anyone who is involved in this study. I further understand that nothing in this consent form is intended to replace any applicable federal, state, or local laws.

Participant's Name (Printed or Typed):

Date:

Participant's Signature or Thumb Impression:

Date:

Principal Investigator's Signature:

Date:

Signature of Person Obtaining Consent:

Date:

Appendix XIV.

Japanese regulatory forms

Form 7

Clinical Trial Notification

Code of test substance	Type of trial	Date of initial notification	Serial number of notification
	1. Company sponsored 2. Physician sponsored		

Name and address of manufacturing site or business office (sample supplier)	
Ingredients and quantities	
Manufacturing method	
Intended indications/efficacy	
Intended dosage and administration	

Outline of trial		
	Objectives	
	Planned number of subjects	
	Target diseases	
	Dosage and administration	
	Trial period	
	Reasons for being onerous	

Name and address of medical institution	Name and title of investigator

Name and title of subinvestigator	Planned quantity of investigational drug	Planned number of subjects	Others (name, etc. of trial applicant in the case of collaborative trial)

Names and titles of coordinating investigator or physicians of the coordinating committee	
Names and titles of contractee entrusted with trial conduct (including duties related to sponsoring) and management of trial, and the scope of the duties	

Remarks	

As indicated above, we hereby submit notification of the clinical trial plan.

Date: _____ / _____ / _____

Address (head office in the case of a corporation)

Name (corporate name and representative name in the case of a corporation)   seal

Independent Administrative Institution, Pharmaceuticals and Medical Devices Agency

Notes:

1. Use Japanese Industrial Standards A4 size paper.

2. In the case of import, enter the name of exporting country, name or corporate name of the manufacturer and trade name in exporting country

3. If all description is not entered in a column, indicate "As stated in the accompanying document ( )" in the column and attach separate documents.

4. Indicate the name and phone/fax numbers of the person to contact and in charge of the notification in the Remarks column.

Form 9

Changes in Clinical Trial Notification

Code of test substance	Type of trial	Date of initial notification	Serial number of notification
	1. Company sponsored 2. Physician sponsored		

Code of test substance	
Notification date Serial number of notification	

Details of changes	Items	Before changes	After changes	Date of changes	Reasons for changes
Remarks					

As indicated above, we hereby submit notification of changes in clinical trial.

Date: _____ / _____ / _____

Address (head office in the case of a corporation)

Name (corporate name and representative name in the case of a corporation)   seal

Independent Administrative Institution, Pharmaceuticals and Medical Devices Agency

Notes:

1. Use Japanese Industrial Standards A4 size paper.

2. If all description is not entered in a column, indicate "As stated in the accompanying document ( )" in the column and attach separate documents.

3. Indicate the name and phone/fax numbers of the person to contact and in charge of the notification in the Remarks column.

Form 11

Clinical Trial Termination Notification

Code of test substance	Type of trial	Date of initial notification	Serial number of notification
	1. Company sponsored 2. Physician sponsored		

Code of test substance	
Notification date Serial number of notification	
Termination date	
Reasons for termination	
Measures taken after termination	

Status of individual medical institution	Name	Quantities of samples supplied (received)	Quantities used	Quantities collected or destroyed	Number of subjects

Remarks	

As indicated above, we hereby submit notification of clinical trial termination.

Date: _____/_____/_____

Address (head office in the case of a corporation)

Name (corporate name and representative name in the case of a corporation)   seal

Independent Administrative Institution, Pharmaceuticals and Medical Devices Agency

Notes:

1. Use Japanese Industrial Standards A4 size paper.

2. If all description is not entered in a column, indicate "As stated in the accompanying document ( )" in the column and attach separate documents.

3. Indicate the name and phone/fax numbers of the person to contact and in charge of the notification in the Remarks column.

Form 13

Clinical Trial Completion Notification

Code of test substance	Type of trial	Date of initial notification	Serial number of notification
	1. Company sponsored 2. Physician sponsored		

Code of test substance	
Notification date Serial number of notification	

Status of individual medical institution	Name	Quantities of samples supplied (received)	Quantities used	Quantities collected or destroyed	Number of subjects

Remarks	

As indicated above, we hereby submit notification of clinical trial completion.

Date: _____/_____/_____

Address (head office in the case of a corporation)

Name (corporate name and representative name in the case of a corporation)   seal

Independent Administrative Institution, Pharmaceuticals and Medical Devices Agency

Notes:

1. Use Japanese Industrial Standards A4 size paper.

2. If all description is not entered in a column, indicate "As stated in the accompanying document ( )" in the column and attach separate documents.

3. Indicate the name and phone/fax numbers of the person to contact and in charge of the notification in the Remarks column.

Form 22 (1) (Related to Article 38 of the Enforcement Regulations of the Pharmaceutical Affairs Law)

			Drugs
Revenue Stamp	Application for approval to manufacture/distribute		Quasi drugs
			Cosmetics

Name	Non-propriety name	
	Brand name	
Ingredients, quantities, or nature		
Manufacturing method		
Dosage and administration		
Indications		
Storage conditions and expiry data		
Specifications and test methods		

Manufacturing factories of product	Name	Address	Category of manufacturing license or accreditation of manufacturer	License or accreditation No.
Manufacturing factories of bulk drug	Name	Address	Category of manufacturing license or accreditation of manufacturer	License or accreditation No.

Remarks	

As indicated above, we hereby apply for approval for manufacture/distribution of (drug, quasi-drug, or cosmetic).

Date: _____ / _____ / _____

Address (head office in the case of a corporation)
Name (corporate name and representative name in the case of a corporation)   seal

Minister of Health, Labor, and Welfare

Governor of Prefectural Government

Notes:

1. Use Japanese Industrial Standards A4 size paper.

2. This form must be submitted in triplicate (original and two copies) or in duplicate (original and copy) to the Minister and Governor, respectively.

3. Entries must be made clearly in black letters using ink, etc.

4. Revenue stamp should be affixed to only the original form and not canceled, except for drugs specified in Article 80-(1)-1 and (2)-5 of the ER-PAL and quasi-drugs specified in the same provisions by the Minister.

5. In the case of an imported drug product of tissue cell origin, enter the name of exporting country, name or corporate name of the manufacturer and trade name in exporting country in the "Manufacturing method" column.

6. If all description is not entered in the "Manufacturing method" column, indicate "As stated in the accompanying document" in the column and attach separate documents.

7. "Storage conditions and expiry data" should be entered only for drugs that require specific storage conditions or expiry date to secure quality.

8. Specifications and test methods are not required for cosmetics.

9. If the product or bulk drug is manufactured by more than one manufacturer, indicate all manufacturers in the column.

10. In the "Category of manufacturing license or accreditation of manufacturer" column, the category specified in Article 26-(1), (3), or (4) or Article 36-(1) or (3) of the ER-PAL must be entered.

11. Proprietors of a pharmacy are required to enter the name of the pharmacy, license No., and license date.

12. Application of products specified in Article 14-(1) according to Article 14-(3)-1 of the PAL must be noted to that effect in the "Remarks" column.

Glossary and abbreviations

A

ADME	Absorption, distribution, metabolism, and excretion
adverse drug reaction	An unintended reaction to a drug taken at doses normally used in humans for prophylaxis, diagnosis, or therapy of disease, or for the modification of physiological function. A causal relationship is at least reasonably possible.
adverse event	A negative experience encountered by a study subject during the course of a clinical trial that is not necessarily associated with the drug. When an adverse event has been determined to be related to the investigational product, it is considered an adverse drug reaction.
amendment	Changes made to a protocol that might significantly affect subject safety, the scope of the investigation, or the scientific quality of the study must be submitted to health authorities in the form of an amendment to the protocol.

B

best practices	Standard methods and work processes considered to represent preferred methods.
biologic	Biologics are any virus, therapeutic serum, toxin, antitoxin, or analogous product applicable to the prevention, treatment or cure of diseases or injuries of humans. In contrast to drugs that are chemically synthesized, biologics are derived from living sources (such as humans, animals, and microorganisms). Most biologics are complex mixtures that are not easily identified or characterized, and many biologics are manufactured using biotechnology.
Biologic License Application	An application to FDA for a license to market a biologic product in the United States.
Biologic License Supplement	An application to FDA to allow a sponsor to make changes in a product that has an approved BLA.
BLA	Biologic License Application
BLS	Biologic License Supplement

blinding	A process used in clinical trials to keep information about the treatments hidden from the subjects and anyone involved with evaluating the subject. Blinding prevents conscious or subconscious biases or expectations from influencing the outcome of the study.

C

Case-control study design	A type of retrospective study design in which subjects with the disease (cases) are compared with subjects who have similar characteristics but who do not have the disease (controls).
CBER	Center for Biologics Evaluation and Research
CDER	Center for Drug Evaluation and Research
CDRH	Center for Devices and Radiological Health
Center for Biologics Evaluation and Research	CBER is the division of FDA responsible for the regulation of biologic products.
Center for Devices and Radiological Health	CDRH is the division of FDA responsible for the regulation of medical devices.
Center for Drug Evaluation and Research	CDER is the division of FDA responsible for regulation of drugs.
CFR	Code of Federal Regulations
CHMP	Committee for Medicinal Products for Human Use
CIP	Clinical investigational plan; see protocol
clinical study report	Final statistical and clinical summary of a given clinical trial protocol. This document was called an integrated clinical and statistical report in the past.
Clinical Trial Authorisation	A European submission requesting permission to test a new product in humans.
Clinical Trial Notification	A Japanese submission requesting permission to test a new product in humans.
CMC	Chemistry, Manufacturing and Control
Code of Federal Regulations	An annual publication that contains the United State's Federal government's regulations. The CFR is divided into 50 titles that represent broad areas subject to Federal regulation. Title 21 is the portion of the CFR that governs food and drugs within the FDA.
cohort	In a clinical study, a well-defined group of subjects who have had a common experience or exposure and are then followed to observe for the occurrence of new diseases or events.

combination product	A product comprising 2 or more individual.regulated components (ie, drug/device, biologic/device, drug/biologic, or drug/device/biologic) that are physically, chemically, or otherwise combined or mixed and produced as a single entity, or 2 or more separate products packaged together in a single package or as a unit, or a product that is packaged separately but is used only with another product.
Common Technical Document	A format agreed on by ICH to organize applications to regulatory authorities to obtain marketing approval.
control group	A comparison group of study subjects who are not treated with the investigational agent. Four types of controls have been defined: no treatment, placebo treatment (no active ingredient), active control treatment (another product), and historical control (data from previous studies or from literature).
CSR	clinical study report
CTA	Clinical Trial Authorisation
CTD	Common Technical Document
CTN	Clinical Trial Notification
curriculum vitae	Compilation of important information for a scientist, particularly one involved in clinical trials. A CV contains education, job history, and publications.
CV	curriculum vitae

D

data listing	Individual data points, also called raw data or line listings
data set	A group of related records that are organized and treated as a unit.
Declaration of Helsinki	A series of guidelines adopted by the World Medical Assembly in Helsinki, Finland in 1964. The Declaration addresses ethical issues for physicians conducting biomedical research involving human subjects.
demographic data, demographics	Subject characteristics, including sex, age, family medical history, and other characteristics relevant to the study in which they are enrolled.
deviation	Changes to the original protocol in the conduct of the study that have not been described in a protocol amendment.

device	An instrument, apparatus, implement, machine, contrivance, implant, in vitro reagent, or other similar or related article, including any component, part or accessory, which is intended for use in the diagnosis, cure, treatment, or prevention of disease. Unlike drugs, devices do not affect metabolism. This characteristic forms the basis for the distinction between drugs and devices, although there are many other characteristics also used in classification.
dosage	Regulated administration of individual doses; usually expressed as a quantity per unit of time
dose	Quantity to be administered at one time or the total quantity to be administered during a specific period.
double-blind study	A study design in which neither the investigator nor the subject knows which medication (or placebo) the subject is receiving
drug	Drugs are chemical entities that affect metabolism, and that are used for treating, curing, or preventing disease in humans or in animals. A drug also may be used for making a medical diagnosis or for restoring, correcting, or modifying physiologic functions

E

EC	European Commission
eCTD	electronic Common Technical Document
efficacy	A product's ability to produce beneficial effects on the duration or course of a disease. Efficacy is measured by evaluating the clinical and statistical results of clinical tests.
EMEA	European Medicines Agency
endpoint	Overall outcome that the protocol is designed to evaluate. Common endpoints are severe toxicity, disease progression, or death. Endpoints are the means by which study objectives are measured.
EU	European Union
European Commission	The European Commission is the board of Europe. The Commission submits bills and supervises the implementation of laws.
European Medicines Agency	A decentralized body of the European Union. Its main responsibility is the protection and promotion of public and animal health, through the evaluation and supervision of medicines for human and veterinary use.
European Union	An economic group currently composed of 27 European nations.
EWG	Expert Working Group
exclusion criteria	Refers to the characteristics that would prevent a subject from participating in a clinical trial, as outlined in the study protocol.

F

FDA	Food and Drug Administration
FD & C Act	Food, Drug & Cosmetic Act
file, filing	Action taken by regulatory authority to permit marketing of a drug or biologic. Often erroneously used when a sponsor is preparing a submission (not preparing a filing).
Food and Drug Administration	Department within the United States Department of Health and Human Services. Enforces Food, Drug and Cosmetics Act and related federal public health laws.
format	A set of electronic file conventions that define the way a document looks and functions. The word is also used to describe the organization of content in a document or submission.

G

GCP	Good Clinical Practices
GHTF	Global Harmonization Task Force
Global Harmonization Task Force	The Global Harmonization Task Force is a voluntary group of representatives from national medical device regulatory authorities (such as the FDA) and the members of the medical device industry whose goal is the standardization of medical device regulation across the world.
GLP	Good Laboratory Practices
Good Clinical Practices	International ethical and scientific quality standard for designing, conducting, monitoring, recording, auditing, analyzing, and reporting studies. Insures that the data reported is credible and accurate, and that subject's rights and confidentiality are protected.
governmental authority	See 'regulatory agency'
GPSP	Good Postmarketing Surveillance Practices
GQP	Good Quality Practices
guidances, guidelines	Guidelines (or guidances as they are currently called) are documents issued by health authorities that represent suggested interpretation of regulations. Unlike regulations, guidelines and guidances are nonbinding recommendations.
GVP	Good Vigilance Practice

H

health authority	See 'regulatory agency'
HED	human-equivalent dose

human-equivalent dose	A dose in humans anticipated to provide the same degree of effect as that observed in animals at a given dose.

I

IB	investigator's brochure
ICF	informed consent form
ICH	International Conference on Harmonisation
IEC	Independent Ethics Committee
IMPD	Investigational Medicinal Product Dossier
IMRaD	Introduction, Methods, Results, and Discussion; layout of a traditional scientific manuscript or journal article
inclusion criteria	Refers to the characteristics that must be met by a subject to participate in a clinical trial, as outlined in the study protocol.
IND	Investigational New Drug application
Independent Ethics Committee	An independent body composed of medical and scientific professionals and nonmedical and nonscientific members, whose responsibility it is to ensure the protection of the rights, safety, and well-being of human subjects involved in a trial and to provide public assurance of that protection, by, among other things, reviewing and approving or providing favorable opinion on, the trial protocol, the suitability of the investigator(s), facilities, and the methods and material to be used in obtaining and documenting informed consent of the trial subjects.
indication	A disease, syndrome, or diagnosis for which the product is intended, and the population for whom it is intended.
informed consent	The voluntary verification of a subject's willingness to participate in a clinical trial, which is accompanied by an informed consent form. This verification is requested only after complete, objective information has been given about the trial, including an explanation of the study's objectives, potential benefits, risks and inconveniences, alternative therapies available, and of the subject's rights and responsibilities in accordance with the current revision of the Declaration of Helsinki.
integrated document	An integrated document synthesizes information from more than one source document.
Integrated Summary of Efficacy	A integrated summary of efficacy results from more than one clinical study, and used for submission to FDA in an NDA.
Integrated Summary of Safety	A integrated summary of safety results from more than one clinical study and used for submission to FDA in an NDA.

Institutional Review Board	An independent group of professionals designated to review and approve the clinical protocol, informed consent forms, study advertisements, and subject brochures, to ensure that the study is safe and effective for human participation. It is also the Board's responsibility to ensure that the study adheres to regulations of a regulatory body such as FDA.
International Conference on Harmonisation	Developed, through a collaboration between the FDA and regulatory agencies in Japan and the European Union, to 'harmonize' regulatory requirements to produce marketing applications acceptable to the United States, Japan, and the countries of the European Union.
Investigational New Drug application	The petition through which a drug sponsor requests the FDA to allow human testing of its drug product
Investigational Medicinal Product Dossier	The basis for approval to conduct clinical trials by the competent authorities in the EU.
investigational product	A medicine, vaccine, or medical device whose quality, safety and/or efficacy are being tested in a specific clinical trial.
investigational site	The physical facility in which a clinical study is conducted. This facility may be a physician's office, a hospital, a medical center, or a clinic.
investigator	A qualified medical professional under whose direction an investigational drug is administered or dispensed. A principal investigator is responsible for the overall conduct of the clinical trial at his/her site.
investigator's brochure	Relevant clinical and nonclinical data compiled on the investigational drug, biologic, or device being studied, which acts as package insert for unapproved products or indications.
IRB	Institutional Review Board
ISE	Integrated Summary of Efficacy
ISS	Integrated Summary of Safety

L

label	The official description of a drug product that includes indication; who should take it; adverse events; instructions for uses in pregnancy, children, and other populations; and safety information for the patient.
line listing	See 'data listing'

M

MAA	Marketing Authorisation Application (Europe) or Marketing Approval Application (Japan)
Marketing Approval Application	Japanese submission requesting approval to market a drug or biologic in Europe.
Marketing Authorisation Application	European submission requesting approval to market a drug or biologic in Europe.
masking	See 'blinding'
maximum tolerated dose	The highest dose of a drug considered to be safe.
Medical Device Directive	Regulatory requirements for conformance with Essential Principles and marketing of medical devices in the European Union
methods	Methods in a clinical study are a description of the study design used, population tested, outcomes (or endpoints) measured, methods of statistical analysis, ethical considerations for protection of the subjects, and operational aspects.
MHLW	Ministry of Health, Labour, and Welfare
Ministry of Health, Labour, and Welfare	The Ministry of Health, Labour and Welfare is one of cabinet level ministries in the Japanese government. This government body provides regulations on maximum residue limits for agricultural chemicals in foods, basic food and drug regulations, standards for foods, and food additives.
MTD	maximum tolerated dose

N

NB	Notified Body
NDA	New Drug Application
New Drug Application	The compilation of all nonclinical, clinical, pharmacologic, pharmacokinetic, and stability information required about a drug by the FDA to approve the drug for marketing.
NOAEL	No-observed-adverse-effect level
No-observed-adverse-effect level	The highest dose of a drug that does not produce a significant increase in adverse effects compared with the control group.
NOEL	No-observed-effect level
No-observed-effect level	The highest dose of a drug that does not produce an effect compared with the control group.
Notified Body	A European private institution that verifies compliance of medical devices (not drugs) in Europe with the applicable Essential Requirements stated in the Medical Device Directive.

O

objectives	Objectives in a clinical study are a statement of the intended purpose of the study, the results of which should support the indication for use. Objectives tend to vary with the phase of development, as early development focuses on safety, and later development tends to focus on efficacy.
outcome	Overall endpoint that the protocol is designed to evaluate. Common outcomes are severe toxicity, disease progression, or death. Outcomes are the means by which study objectives are measured.

P

PAB	Pharmaceutical Affairs Bureau
package insert	Printed data sheet given to subject that provides information on safety, dosing, side effects, etc of the prescribed drug. It is usually multiple pages and is folded inside the carton that contains the prescription medicine.
PAL	Pharmaceutical Affairs Law
parallel enrollment	The enrollment of study groups at the same time, in contrast to enrollment of one group after another.
PFSB	Pharmaceutical and Food Safety Bureau
pharmaceutical	See 'drug'
Pharmaceuticals and Medical Device Agency	An independent administrative Japanese institution, which does the work of the Japanese regulatory body, the MHLW. PMDA does not issue laws or notifications, but is responsible for reviewing submissions for drugs, biologics, and medical devices in Japan.
pharmacodynamics	The study of the biochemical and physicological effects of drugs, and the mechanisms of drug action and the relationship between drug concentration and effect.
pharmacokinetics	The study of absorption, distribution, metabolism and excretion (ADME) of bioactive compounds in a higher organism.
phase	A successive set of steps in clinical development. Phases in a clinical study influence many design features, such as the number of subjects and the frequency and type of measurements.
placebo	An inactive substance designed to resemble the drug being tested. It is used as a control to rule out any psychological effects testing may present. Most well-designed studies include a control group that is receiving a placebo.

PMA	Premarket Approval Application
PMDA	Pharmaceuticals and Medical Device Agency
PMSB	Pharmaceutical and Medicinal Safety Bureau
pooling	The combining of several databases that are analyzed as a single database
postmarketing	Period after a drug or device has received marketing approval from a regulatory body.
prospective enrollment	The collection of data going forward in time, in contrast to analysis of data that already exist and are studied under designs such as a case-control or historical control.
protocol	A study plan on which all clinical trials are based. The plan is carefully designed to safeguard the health of the participants as well as answer specific research questions. A protocol describes what types of people may participate in the trial; the schedule of tests, procedures, medications, and dosages; and the length of the study.
PSUR	Periodic Safety Update reports

R

randomization	A method based on chance by which study participants are assigned to a treatment group. Randomization minimizes the differences among groups by equally distributing people with particular characteristics among all the trial groups.
randomized clinical trial	Study design considered to be the gold standard. In a randomized clinical trial, a control is used to compare the results of treatments.
regulations	Rules prepared by government agencies and used to administer a law.
RCT	randomized clinical trial
regulatory agency	Governmental authority responsible for granting marketing approval for drugs, biologics, devices, and combination products.
RTF letter	Refusal to file letter
retrospective study	A study that looks backward in time, usually using medical records and interviews with patients who are known to have a disease.

S

safety	An evaluation of whether the product may cause toxic or harmful effects; safety is assessed in every clinical study, irrespective of developmental phase or product classification.

safety monitoring committee	Sometimes called a data monitoring committee. Group of individuals with pertinent expertise that reviews on a regular basis accumulating data from an ongoing clinical trial. This group advises the sponsor regarding the continuing safety of current participants and the continuing validity and scientific merit of the trial. The committee can stop a trial if it finds toxicities or if treatment is proved beneficial.
serious adverse event	A serious adverse event is any adverse event that results in death, is life threatening, requires inpatient hospitalization, is persistent or causes significant disability/incapacity, or causes congenital anomaly or birth defect.
Shonin	A document approving a Japanese Marketing Approval Application, and allowing marketing and distribution of a new product in Japan.
single-blind study	A study design in which one party, either the investigator or subject, is unaware of what medication (or placebo) the subject is taking
SMPC	Summary of Product Characteristics
source document	In regulatory writing, a source document records information relevant to one clinical study, and forms the basis for all integrated documents and submissions, which recorded information from more than one clinical study. A source document in clinical research is a document in which data collected for a clinical trial are first recorded, generally the subject's medical record.
sponsor	Individual, company, institution or organization taking responsibility for initiation, management and financing of study.
statistical power	Statistical power is the probability you will detect a meaningful difference, or effect, if one were to occur.
stopping rules	Established safety criterion that would either pause or halt the study because of futility or risk(s) to the participants in or to be enrolled in the study.
study	Also called a trial; it is the conduct of the protocol.
subject	An individual participating in a research study. A subject may be a healthy individual, or a person with the disease or condition under investigation.
submission	Documentation provided by a sponsor to seek approval for the marketing of a drug or biologic.

T

third-party blinding   When blinding of the subject and the investigator is not possible
                            (such as with implantable medical devices, because, of course, pla-
                            cebo devices are not a possibility) a third party responsible for
                            observing study results, such as laboratory values or radio-imag-
                            ing scans, may be blinded. A third-party blind study design also
                            is used when unintentional unblinding may occur. Unintentional
                            unblinding occurs when a clinical sign (rash, a decrease in blood
                            pressure) allows the subject or investigator to guess the identity
                            of the treatment.

toxicity   Some of the possible side effects of a drug; also indicates how
                            much of a drug can safely be taken.

treatment compliance   Each subject's adherence to protocol-specified procedures.

V

volunteer   A healthy person who agrees to be part of a clinical trial; may
                            receive active study drug or placebo or other treatments

W

WHO   World Health Organisation

WMA   World Medical Association

Index

Numbers in *italics* refer to figures and tables. Numbers in **bold** refer to Appendices.

Clinical Trial Registries

A Practical Guide for Sponsors and Researchers
of Medicinal Products

Foote, MaryAnn (Ed)

Foote, M.A. (Ed)
Clinical Trial Registries
2006. XIII, 194 p. 8 illus.
Softcover
EUR 32.00 / CHF 48.00
ISBN 978-3-7643-7578-2

Clinical Trial Registries: A Practical Guide for Sponsors and Researchers of Medicinal Products is a necessary addition to the library of all researchers who plan to publish their results in top-tier, peer-reviewed journals. ICMJE editors and other journal editors require registration of clinical trial information on publicly available Web sites before enrollment of study subjects and some countries and regions also require this information, as well as timely publication of study results. Not only does this book discuss the genesis of these requirements, it also provides practical information for researchers and sponsors on how to establish a workflow for a clinical registry project, how to file to a registry, and how to post results. More than 25 current Web addresses for registries are provided as well as a comprehensive annotated bibliography of papers on the topic of clinical trial registries. This book is a valuable source of information for all sponsors of medicinal products.

From the Contents:
Clinical trial registries and publication of results.- The journal editor's perspective.- Industry perspective on public clinical trial registries and results databases.- Public and patient usage and expectations for clinical trial registries.- Building a global culture of trial registration.- The Japanese perspective on registries and a review of clinical trial process in Japan.- Transparency and validity of pharmaceutical research.- A project management approach to the planning and execution of clinical trial registries.- Biopharmaceutical companies tackle clinical trial transparency.- In search of 'Clinical Trial Register – Version 2.0'.- Clinical trial registries and study results databases.- Annotated bibliography of important papers

Numerous practical hints for everyone involved in registration of clinical projects Useful Web-addresses Different countries' regulations are considered

www.birkhauser.ch

Lightning Source UK Ltd.
Milton Keynes UK
03 June 2010

155037UK00001B/12/P